An Economic History of England
Edited by T. S. ASHTON

The 18th Century

An Economic History of England
Edited by T. S. ASHTON

E. *Carus-Wilson:* THE MEDIEVAL PERIOD
F. J. *Fisher:* THE 16TH AND 17TH CENTURIES
T. S. *Ashton:* THE 18TH CENTURY
A. H. *John:* 1800-1875
W. *Ashworth:* 1870-1939

An Economic History of England: The 18th Century

by

T. S. ASHTON

*Professor Emeritus of Economic History in the
University of London, London School of Economics*

METHUEN & CO. LTD, LONDON
11 New Fetter Lane, EC4

First published April 4th 1955
Reprinted with minor corrections 1955
Reprinted 1959
Reprinted with minor corrections four times
Reprinted 1969

Printed and bound in Great Britain by
The Camelot Press Ltd., London and Southampton

SBN 416 57340 1
1.7

Distributed in the U.S.A.
by Barnes & Noble Inc.

Preface

THE five volumes that are to make up this History will offer a survey of the centuries from the early Middle Ages to our own day. The area concerned is England: references to Scotland, Ireland, Wales, and other countries are introduced only for purposes of comparison. The authors are not professional economists, but their object is to find answers (partial and provisional though these must be) to the questions economists ask, or should ask, of the past. If, being human, they turn now and then to wider issues their observations on these are incidental to the main purpose. And if some topics admittedly of an economic nature are treated lightly, that is in some cases because our knowledge is limited, but in others because they have been dealt with adequately elsewhere. The object is to supplement and extend, rather than to supplant, existing, well-known treatises.

All historical writing that is worth while must be individual: no attempt is made at uniformity of treatment. But the volumes are the product, not only of personal research, but also of discussion in seminars and of gossip in common rooms and corridors. They may be thought of as representing the way in which the subject has been taught, in recent years, at the London School of Economics and Political Science.

This first contribution to the series will disappoint many readers. I have deliberately laid stress on the continuity of economic life in the eighteenth century and have said little of technological change, of the policies and ideas of the period, or of modern reconstructions of these enshrined in such words as Capitalism, Mercantilism, and Imperialism. I confess to distaste for these imprecise terms, which seem to me to have blurred, rather than sharpened, our vision of the past. A few months ago, indeed, in an Oxford common room, I boasted that no single word ending in 'ism' would be found in this book. 'Not even "baptism"?' asked one of my hosts, in that gentle way they have of deflating you at Oxford. The proofs had not yet been returned to the publisher: it would have been easy to substitute 'christening'; but I decided to leave the offensive syllables as a warning to myself against vainglory.

One of the chief attractions of the School of Economics is the encouragement it gives to specialised investigation. For seven or eight years the Research Committee has put at my disposal the services of research assistants whose chief task has been to assemble statistical material relating to the period covered by this book. I owe much to the energy and enterprise, in succession, of Derek Froome, Cyril Ehrlich, Walter Elkan and Andrew Templeton. Only a small part of their findings is presented in the Appendix. It is offered, not merely as evidence for statements in the text, but as a means of enabling readers to satisfy their curiosity on other matters, and perhaps to discover for themselves relationships I have failed to notice. Some of the tables are printed here for the first time. I thank H.M. Commissioners of Customs and Excise for permission to reproduce them.

The early part of the book was written when I was the guest of Johns Hopkins University in 1952. I owe much to the facilities given me to work on the impressive collection of pamphlets in the Hutzler Library there, and still more to the stimulus I received, in those gracious precincts, from conversations with G. H. Evans, Fritz Machlup and E. D. Domar, among others. But my chief debt is to my colleagues at the School of Economics and especially to F. J. Fisher and A. H. John, who have generously provided me with the fruits of their wide reading and reflections, and have read through the work in proof. In my absence abroad, Dr John undertook also the preparation for the press of the statistical material, and, throughout, he has relieved me of much tiresome work in checking references. Finally, I am deeply grateful to Mrs Gladys Cornwell, whose skill in typing, and tact in deflecting would-be intruders on the limited time I had for writing, are beyond all praise.

T. S. ASHTON

The London School of Economics,
December, 1954

Contents

	PREFACE	v
I	THE PEOPLE OF ENGLAND	1
II	AGRICULTURE AND ITS PRODUCTS	30
III	INTERNAL TRADE AND TRANSPORT	63
IV	MANUFACTURES	91
V	OVERSEAS TRADE AND SHIPPING	130
VI	MONEY, BANKING AND FOREIGN EXCHANGE	167
VII	LABOUR	201
	APPENDIX—STATISTICAL TABLES	237
	INDEX	255

CHAPTER I

The People of England

I

THE eighteenth century was the age of political arithmetic. It is true that it was the age of much more than this: of reason and toleration, Newtonian cosmography, Palladian architecture, wit, good taste, and good manners. But, viewed through the narrow window of the economic historian, it was an age when interests were directed largely to things that could be measured and weighed and calculated. If men may be judged by their utterances, the educated no longer troubled their minds over-much with high matters of doctrine and polity: they were exercised less with the purpose of life than with the art of getting a living, less with the nature of the state than with the means of increasing its opulence. In 1700 there were fewer men searching the Scriptures and bearing arms than there had been fifty years earlier, and more men bent over ledgers and busying themselves with cargoes. There were fewer prophets and more projectors, fewer saints and more political economists. And these economists were concerned less with principles of universal application than with precepts derived from experience, less with what was ultimately to be wished for than with what was immediately expedient.

There was a demand for precise information in numerical form. This was satisfied partly by drawing on sources that had long been available: parish registers, bills of mortality, hearth-tax and poor-law returns, among others. But, since 1688, fresh springs of detailed knowledge had appeared in reports on public revenue and expenditure, trade and navigation, production, the state of the coinage, and the prices of foodstuffs and other commodities. Figures formed the hard core not only of Parliamentary debate, but also of the treatises, pamphlets, newspaper articles, and broadsheets by which public opinion was shaped. An economic history of the period cannot be other than statistical in substance. It need not, it may be added, for that reason, be loaded with tables and disquisitions on methods. Contemporaries themselves did most of the compilation and criticism of sources: the reader who seeks detail can find it, for the most part, in printed, accessible works.

There was no matter about which people were more inquisitive than that of their own numbers. A few years before the century opened Gregory King had used the hearth-tax returns to make an estimate which, according to Davenant, was 'perhaps more to be relied on than any thing that had ever been done of the like kind'.[1] Close scrutiny by a distinguished demographer[2] supports Davenant's judgment: King's reckoning that the people of England and Wales numbered some $5\frac{1}{2}$ millions in 1695 is probably as close to the truth as the nature of the evidence allows. In the early 1750's controversy arose as to the naturalisation of aliens, and this led to a protracted debate as to whether population was increasing or declining. The protagonists were mainly clergymen and ministers of religion; and the discussion generated heat rather than light. Proposals to settle the matter by taking a census or collecting vital statistics encountered strong opposition on the varied grounds that such measures would incur divine displeasure, be subversive of liberty, and afford information that might be of use to the enemies of Britain.[3] It was not until after the Rev. Thomas Malthus had published his powerful and alarming *Essay* that an enumeration of the people was made; and not until much later that the public submitted to compulsory notification of births, marriages, and deaths. The total of 8,872,000 people recorded by the census of 1801 was deficient, because of the omission of soldiers and marines. But the revised estimate of 9,168,000, made by John Rickman, cannot be far from the mark.[4] It is fairly safe to say that between 1695 and 1801 the population of England and Wales had increased by about two-thirds.

Such a growth of numbers was new to English experience. Exactly how and why it happened has never been firmly established. But informed opinion lays stress on two factors: the disappearance, after

---

[1] Gregory King, *Natural and Political Observations upon the State and Condition of England*. Appended to George Chalmers, *An Estimate of the Comparative Strength of Great Britain* (1802 ed.), p. 412.

[2] D. V. Glass, 'Gregory King and the population of England and Wales at the end of the Seventeenth Century', *Eugenics Review*, vol. xxxvii, No. 4, pp. 170-83; 'Gregory King's estimate of the population of England and Wales, 1695', *Population Studies*, March, 1950.

[3] D. V. Glass, 'The population controversy in eighteenth-century England. Pt. I, the background', *Population Studies*, July 1952.

[4] *Report of the Census (1811) Preliminary Observations*, p. xx.

1665, of the plague that had afflicted Christendom for centuries,[1] and the lowered incidence of famine and the diseases resulting from insufficiency of food.

Any attempt to trace the course of population between 1695 and 1801 must be based on information drawn from the registers in which the parish clergy recorded the baptisms, marriages, and burials at which they officiated, and the bills of mortality compiled by lay agencies for London and other large towns. These are far from complete. Baptisms are not synonymous with births, or burials with deaths. Most dissenters, and some adherents of the Church, were unwilling, or unable, to pay the fees required by the clergy; and, as nonconformity spread, the registers must have become increasingly defective as a basis for vital statistics. This has not prevented demographers from using them to frame estimates of the population for decennial periods between 1700 and 1780, and for each year between 1780 and 1800. The historian must admire the ingenuity and skill that went to the making of these. But he cannot accept the assumptions on which they are based as though these were statements of reality; and for this reason little reference will be made to the substantial body of material assembled in several statistical works.[2]

Nevertheless one or two observations can be made as to the way in which population grew in this period. The parish registers and bills of mortality suffice as evidence that the expansion did not result from a steady increase of births or a steady decline of deaths: there were wide annual variations of each. They suggest, further, that growth was slow until about 1750, that the pace increased during the following three decades, and that it was rapid in the last twenty years of the century.

It is not difficult to suggest reasons why there should have been little advance in numbers before 1750. The bills of mortality show that burials were more numerous in the early, than in the later months of the year, and that they varied considerably with the degree of severity of the winter. Now it is true that there was in the

---

[1] For one theory of the cause of this see H. Zinsser, *Rats, Lice and History* (1936).

[2] For the estimates made by John Rickman, see the *Enumeration Abstract* for England and Wales (1843), Preface, pp. 34-7. For a clear statement of the assumptions on which these and later estimates were based, see E. C. K. Gonner, 'The Population of England in the Eighteenth Century', *Journal of the Royal Statistical Society*, vol. LXXVI, Pt. III, pp. 262-303, and G. Talbot Griffith, *Population Problems in the Age of Malthus*, chs. I and II.

first half of the eighteenth century no such a run of severe frosts and other adverse circumstances as in the 'barren years', 1692-9. But there were three periods (1708-10, 1725-9, and 1739-42) when excessive cold or rain led to a sharp increase in the number of burials. In the first and third of these large numbers of people perished of cold or of fevers engendered by this, and in all three many others died of hunger or of ailments resulting from shortage of food. Each dearth was accompanied or followed not only by an increase in burials, but also by a fall in baptisms.[1] For, not only were many potential parents swept away, but poverty and distress, it may be supposed, led to some postponement of marriage. A lower number of births in the years following 1709 meant a lower number of young adults twenty or so years later: it may well explain the fall in marriages that students of the registers observe in the 'thirties.

Even in normal years mortality was, by modern standards, deplorably high. Nothing varies more than the nomenclature of disease, and some of the terms used in the bills of mortality are imprecise. 'Fever' seems to have denoted any complaint that led to a rise of temperature; and 'lethargy', which does not sound highly unpleasant, covered a range of terrifying ailments. Little ambiguity, however, attaches to 'diarrhoea', 'smallpox', and 'consumption' which designated three of the maladies responsible for the high incidence of death among the young. Of the total burials recorded in London in 1739 slightly more than a half were those of children under eleven, and 38 per cent. were of infants under three years of age. (In other years even higher percentages were recorded.) London was, no doubt, exceptionally inhospitable to the young, but there is evidence that

[1] The following figures extracted from the London bills of mortality are for annual periods beginning in October and ending in September:

	Burials	Baptisms
1707-8	21,270	16,120
1708-9	22,100	15,520
1709-10	24,710	14,880
1710-11	20,510	14,630
1724-5	25,730	18,840
1725-6	29,210	19,140
1726-7	28,210	18,450
1727-8	28,620	17,290
1728-9	27,920	16,800
1738-9	26,320	16,420
1739-40	30,920	15,540
1740-1	28,450	14,850
1741-2	31,590	13,760
1742-3	25,090	15,020

in all parts of the country the expectation of life of a newly born infant was small. In these circumstances only an extremely high birth-rate could have resulted in even a moderate increase of population.

There is, however, no evidence that the birth-rate was inordinately high. If, as several writers pointed out, religious zeal rarely led to celibacy in England, as it did in France, custom and the operation of the Poor Law may have had a similar, if less marked, effect. Servants in husbandry were still accustomed to live with their masters in the farmhouses, instead of in homes of their own; and apprentices were usually prohibited from marrying until they had served the full term of years required by their indentures. Regional attachments set bounds to the marriage market. According to Defoe, the gentlemen of Cornwall seldom went outside the county for a wife or the ladies for a husband, 'from whence they say, that proverb upon them was rais'd (viz) That all the Cornish gentlemen are cousins'.[1] It has been suggested[2] that in rural areas a poor man would not often look for a wife outside his own neighbourhood; and since in any one parish or hundred the number of the two sexes would rarely be exactly equal, there would be some men or women for whom there were no partners. This was certainly the case in some of the towns. Defoe mentions that at Eltham in Kent 'there was abundance of ladies of very good fortunes . . . but 'tis complain'd of that the youths of these families where those beauties grow, are so generally or universally bred abroad, . . . that for ladies to live at Eltham is, as it were, to live recluse and out of sight; since to be kept where the gentlemen do not come, is all one to be kept where they cannot come'.[3] Later in the century a number of local censuses were taken. According to an enumeration made in January, 1790, the parish of Tiverton contained 1,776 men and 2,387 women.[4]

Class distinctions, also, sometimes imposed celibacy: women, in particular, might remain single, not because of a shortage of men, but because the candidates for matrimony were socially ineligible. Preston—'Proud Preston'—provides an illustration. 'This town subsists,' it was said, '. . . by many families of middling fortune who

---

[1] D. Defoe, *A Tour through England and Wales* (Everyman ed.), vol. I, p. 234.
[2] By Mr. D. Eversley of the University of Birmingham.
[3] D. Defoe, *op. cit.*, vol. I, p. 100.
[4] M. Dunsford, *Historical Memoir of the Town and Parish of Tiverton* (1790) p. 464.

live in it, and it is remarkable for old maids, because their families will not ally with tradesmen, and have not sufficient fortunes for gentlemen.'[1] Nor was such a tendency peculiar to the upper and middle classes. Few Scottish colliers could persuade women other than the daughters of colliers to wed with them, and there may have been similar endogamous groups among the workers south of the Tweed. It is not to be believed that large numbers of Englishmen were so lacking in spirit as to allow geographical, administrative, or class barriers to deter them from matrimony, or that alliances outside wedlock were few. But it is not fanciful to suggest that the existence of non-competing areas in society may have had some effect on marriages, and hence on births, at this time.

The margin between births and deaths must in any case have been relatively small. But it was narrowed by the spread of pernicious ways of living and especially by an excessive indulgence in spirits. During the wars of William III British 'geneva' and whisky had come into use as substitutes for French brandy. These cost little to produce and sold at a low price. Since the distillers offered a market for home-produced wheat and barley, they were looked on with favour by the landlords, and their industry was encouraged by Parliament. For more than a generation the only restraint on production was a low excise duty, of from 3*d*. to 6*d*. a gallon, and retail trade in spirits was open, without licence, to all. The result was a steep rise in the number of deaths and a demoralisation, of rich and poor alike, so great as to lead to fears not only for the future of civilised life in Britain, but even for the continuance of the race.

Early attempts to deal with the problem proved futile. An Act of 1736[2] recites that 'the drinking of spirituous liquors or strong waters is become very common, especially among the people of lower and inferior rank, the constant and excessive use whereof tends greatly to the destruction of their healths, rendering them unfit for useful labours and business, debauching their morals, and inciting them to perpetrate all manner of vices; and the ill consequences of the excessive use of such liquors are not confined to the present generation, but extend to future ages, and tend to the devastation and ruin of this kingdom'.

[1] R. B. Westerfield, *Middlemen in English Business*. p. 200, quoting R. Pococke, *Travels through England*, Camden Society, vol. I, p. 12.
[2] 9 Geo. 11, c. 23.

The People of England

The imposition by this act of a penal tax on both distillers and retailers was followed by a temporary fall of output. The chief result, however, was to drive the trade underground. Gin could still be obtained from tallow chandlers and others who housed illicit dealers 'hid behind some wainscot . . . or otherwise concealed'. It could be bought on the streets. Under an Act of 1738 an offender who failed to pay the fine of £10 for illegal hawking of spirits was to be committed to the House of Correction for two months, and 'before his or her discharge therefrom, be stript naked from the middle upwards, and whipt till his or her body be bloody'. But measures like this served only to intensify opposition to regulation: it must have needed courage to serve as an exciseman in London or Bristol at this time.

In 1742-3 the output of British spirits reached its peak of over 8 million gallons, a figure roughly six times as great as that of the early years of the century. Perhaps because of the repeal[1] of earlier Acts, and the adoption of a more moderate policy in this year, the upward course of production was halted. But it was not until 1751, when distillers were forbidden to sell by retail, and when both spirit and licence duties were again sharply raised,[2] that the gin age came to an end.[3]

It was in London and the other large towns that addiction to gin was most serious, but there is evidence that the country districts were also affected. As in the case of other malevolent forces, widespread consumption of spirits was reflected not only in a high rate of deaths, but also in a low rate of births. Contemporaries held—and the belief is not to be ruled out—that heavy drinking of gin was inimical to fertility. If modern science were to confirm the opinion there would be no need to look for further explanation of the retarded growth, or actual decline, of population that seems to have taken place in the 'thirties and 'forties.

Harsh climatic conditions, dearths, diseases, social immobility, and alcoholic excess may suffice, then, to explain the slow pace of advance of population in the first half of the eighteenth century. After 1750 some of the depressive forces were weakened. In several years harvests were poor, but, until 1800, there was nothing approaching a

---

[1] 16 Geo. II, c. 8.   [2] 24 Geo. II, c. 11.
[3] For a graphic account of the effects of spirit-drinking at this period see M. D. George, *London Life in the Eighteenth Century*, pp. 27-42, from which the quotations in the two preceding paragraphs are drawn.

famine. Technical improvements in agriculture increased the output of grain and cattle. Innovations in transport made it possible to meet local shortages by bringing supplies from other areas. The growth of overseas trade enabled a national deficiency to be overcome by drawing grain from overseas. And better facilities for storage led to the holding of larger reserves. 'Such a run of wet seasons as we have had the last ten or twelve years would have produced a famine a century or more ago.'[1] In 1773, when Gilbert White wrote these words, life was becoming a little less precarious than it had been.

There was more variety of diet: a greater use of vegetables did something to reduce scurvy. Larger supplies of coal made cottages less damp and ill-ventilated, and more soap and washable garments led to improved hygiene. Under the influence of men (largely Scots) trained by pupils of Boerhaave of Leyden and von Haller of Göttingen, medical practice improved; and a new era in surgery resulted from the work of the two brothers, William and John Hunter. Inoculation against smallpox had been brought to England from Constantinople as early as 1718. When, after the middle of the century, its practice spread from the aristocracy to the poor, one instrument of death lost some of its potency. Hospitals were established in several of the larger towns, and dispensaries were set up for the treatment of poor people as out-patients.

It is true that the beneficial effects of such developments were partially offset. The period was one of large-scale warfare, and casualties in the armed forces, and the epidemics that war brings in its train, tended to increase the number of deaths. Towns were growing. Professor Heckscher[2] quotes a writer of 1741 who declared of Sweden that 'the secret damage done to the state through the towns must almost be considered equivalent to a pestilence'. In the parish registers of large towns in England the entries relating to burials still normally outnumber those relating to baptisms. Before the coming of sanitary science concentration of people was almost always hostile to health and longevity.

On the other hand, there is a good deal of evidence that births were increasing. It may be that genetic forces of which little is yet known were responsible for this. But there is no necessity to assume any rise of fertility, since changes that were taking place in the

[1] Cited by T. H. Baker, *Records of the Seasons*, Preface, p. vi.
[2] E. F. Heckscher, 'Swedish Population Trends before the Industrial Revolution', *Ec. Hist. Rev.* Second Series, vol. II, No. 3 (1950), pp. 266-77.

of the country. In the late seventeenth century there had been a large influx of French and Dutch, some expelled from their mother land by religious intolerance, but others drawn to England by economic considerations. In the eighteenth century the stream of immigrants continued. It included Germans who had fought in the wars of Anne, persecuted Jews from central Europe, and Moravians and others in whom the zeal of the proselyte and the desire of the natural man for advancement were blended in varying proportions. (There must have been many who—as was said of the American missionaries to Hawaii—came to do good and did well.) Some of the new industries that were springing up in the early years of the century (paper-making, glass-making, and silk-throwing, for example) owed much to immigrants from Europe; and the list of names of merchants who presented their congratulations to George III on his accession is eloquent of the part played by men of continental origin in the commerce of London. More important than any of these, however, were the Scots who streamed south, bringing stout arms and trained minds to the service of English industry, and of the Irish who, in increasing numbers, crossed the sea annually to help to gather the harvests, or to settle in London as market porters, builders' labourers, or coal-heavers. No valid statistics of immigration are available. In the war of pamphlets as to the desirability of admitting aliens some extreme estimates were put forth. A writer of 1752 asserted that more than half the male inhabitants of London were Catholic Irish—and added that 'if any guess can be made from the number of executions they are certainly more than half the bad ones'.[1] Hangings at Tyburn are, however, a poor form of statistical evidence, for in the law courts the dice may have been loaded against the immigrant Catholic. More precise evidence is offered by figures collected at the Westminster General Dispensary, and cited by Dr. M. D. George. Of the married people treated between 1774 and 1781, 1 in 11 had been born in Ireland, 1 in 15 in Scotland, and 1 in 60 outside the British Isles.[2] But London attracted more immigrants than most other places, and it is not possible to draw inferences from these figures as to the proportions of Irish, Scots, and foreigners in the country as a whole.

[1] *Reflections on various subjects relating to Arts and Commerce* . . . (1752), p. 83.
[2] M. D. George, *op. cit.*, p. 111. Dr. George points out that if a similar investigation had been made in other parts of the capital it would have shown a higher proportion of Irish.

structure of society would seem to offer sufficient explanation of a rise in the number of births. As agricultural holdings became larger the proportion of the labourers who had cottages and families of their own must have risen. As apprenticeship weakened, the average age of marriage must almost certainly have fallen, and the average number of children to a family have increased. Of at least equal importance was the widening of the area within which men and women looked for their partners in matrimony. This seems to have had an effect on the quality of the people; for, as has frequently been pointed out, with a decline in the marriages of near relatives, the number of village idiots was reduced. But it must also have had an effect on the size of the population. It is a simple statistical principle that the larger the group the smaller proportionately will be the surplus of one sex over the other; for within a wide region an excess of women in any one part is likely to be offset by an excess of men in another. As the larger rural area came to take the place of the parish as the effective unit of social life, marriages in the country districts probably increased. And as the towns, with their fairs and other attractions, drew in young people (whether as residents visitors) match-making may well have been facilitated. It has long been recognised that a growth of the regional, as distinct from the parochial or village, economy had a stimulating effect on industry output: it had almost certainly a similar fructifying effect population.

Of the children born a larger proportion lived to maturity. numbers of those who died shortly after birth, in the British Ly in Hospital, fell progressively from 1 in 15 in 1749-59 to 1 in in 1799-1800.[1] In other parts of the country there can hardly been such spectacular improvement. But both in London and provinces there is evidence of a decline in the virulence of maladies as smallpox and dysentery—whether because of impr medical practice or because of better and more abundant su of food and houses. Anything that reduces juvenile mortality have cumulative effects on population. For it preserves not individuals, but also potential parents. The acceleration of nu in the second half, and especially in the last two decades, eighteenth century was probably due more to a fall of infant juvenile deaths than to any prolongation of adult life.[2]

Account must be taken of the movement of people into

[1] G. Talbot Griffith, *op. cit.*, p. 241.   [2] *Ibid.*, pp. 209-1

The People of England

Immigration was offset, at least in part, by a movement of Englishmen overseas. Some went as indentured servants or redemptioners to Maryland and Virginia, and large numbers of convicts were shipped to these and other colonies. Some—generally men of higher station in life—left their homes to become planters or agents in the West Indies; and yet others took service with His Majesty or with the Honourable East India Company, sailed in far parts of the world, and never saw England again. All the time there was a trickle of skilled artisans to France, the Low Countries, Russia, New England, and other places where attempts were being made to set up new industries based on British techniques. In 1695 Gregory King estimated that the Plantations took a thousand people 'over and above the accession of foreigners' to England. But, with so many ports in the British Isles and so much coming and going between them, it was impossible for the state to keep itself informed of the movement of its subjects, and there could be no reliable statistics of either immigration or emigration. The balance probably swung from side to side according to the state of employment at home: there is some evidence of abnormal exodus in years when industry and trade were depressed.[1] Over the century as a whole, however, English manufactures were offering increased opportunities to labour at rates of pay well above those of most other countries, and for this reason we may guess (though we cannot affirm) that the men and women who moved out of the country were outnumbered by those who came in from other parts of the British Isles and abroad.

Evidence drawn from the parish registers and the hearth-tax returns suggests that in 1700 the centre of population lay well to the south. There were concentrations in and about London and in the wide region of which Bristol and Exeter were the chief towns.[2] But

---

[1] As, for example, in 1774. *The Leeds Mercury* of May 17th, 1774, reports that 'scarce a week passed without some setting off from this part of Yorkshire for the Plantations' on account of bad trade and the high cost of living. Cited in *Thoresby Miscellany*, vol. XLI, Pt. 2, p. 175.

[2] It has been calculated that seven counties in these two areas held a third of the population of England. E. C. K. Gonner, *loc. cit.* The counties are Middlesex, Surrey, Kent, Gloucestershire, Somerset, Wiltshire, and Devonshire. Professor Gonner's calculations were based on the figures of Rickman and are therefore to some extent open to question. He suggests, however (p. 286), that they can be used to indicate the distribution of population 'since a similarity of error, if error there be, may be assumed uniformly.' This is true of the error that may arise from assuming a stable relation over time between baptisms, marriages, and burials on the one hand and population on the other. It does not dispose of the error that may come of assuming a fixed proportion of nonconformists throughout the country.

there were also areas of considerable density in the east (Norfolk, Suffolk, and Essex) and the north-west (Lancashire and the West Riding). It has been said that people tended to congregate along the coasts and that the inland parts were sparsely populated.[1] But Northumberland, Yorkshire (excluding the West Riding), Lincolnshire, Sussex, Hampshire, and the whole of Wales are known to have been sparsely populated.[2] Yet all these had long stretches of coast. The majority of the English and Welsh were agriculturists, and the most fertile, and hence best populated, of the agrarian counties were those of the midland plain, far away from salt water. It is true that proximity to the sea was an important factor in the growth of communities of coal miners in the north-east and of tin and copper miners in the south-west. It is true also that at London, Bristol, and other large ports there were concentrations of workers engaged not only in shipping and shipbuilding, but in trades like milling, brewing, distilling, tanning, and sugar boiling, which were carried on in fairly large establishments and called for the bulk transport that only the sea could provide. But the people employed in such activities must have been far outnumbered by those who worked on lighter materials in their own homes; and most of these were in the hinterland. The largest industry of all—the manufacture of cloth—rarely flourished in places near to the sea. Large stretches of the English coastal strip consist of limestone, and the woollen industry could not tolerate hard water. Political sentiment was opposed to the settlement of spinners and weavers near to the seaboard: it was difficult enough, as things were, to enforce the law against the exportation of wool. For this reason alone, a scheme to plant weavers along Cardigan Bay, and another to provide husbands for maidens skilled in spinning, and to settle them in the underpeopled area from Kent to Dorset, were bound to come to naught. And for reasons connected with supplies of fuel and with the structure of industry, many crafts, beside the making of woollens, shunned the coasts. The manufacture of hardware, nails, cutlery, pottery, linen, cotton, and hosiery was carried on by men and women most of whom can never so much as have

---

[1] Writing of the period of Defoe's *Tours* in the seventeen-twenties, Professor Cole asserts that 'save for the widely spread provenance of the wool industry, manufacturers avoided, wherever possible, the interior of the country, and gathered closely round the ports and the navigable rivers'. Defoe, *op. cit.*, p. xi.

[2] For a map showing the density of population per square mile see E. C. K. Gonner, *loc. cit.*, p. 289.

set eyes on the sea. The truth seems to be that, in the early part of the century, both agricultural and manufacturing workers were to be found in coastal and inland areas alike, but that the balance almost certainly inclined to the interior. What is beyond dispute is that at this time people were spread over the face of the country more evenly than at any time since.

As time went on the demographic centre moved northward. The iron industry, starved of charcoal, shifted from Sussex and Gloucestershire to Shropshire, Worcestershire, Staffordshire, and South Yorkshire. East Anglia lost ground to the West Riding in the manufacture of worsteds. From London the making of firearms was transferred to Birmingham; silk weaving to Coventry and Cheshire; hosiery manufacture to Nottingham, Derby, and Leicester; and (towards the end of the century) calico-printing to Lancashire. It seems probable that the change in the demographic map was the result not of differences in 'natural increase' between regions, but of industrial migration. 'Manufactures, by their own natural course, will ever remove from the dearer to the cheaper places', said a writer in 1752; and even at this relatively early date he was able to add 'ours evidently tend Northward'.[1] Sometimes when a firm moved the workers moved with it. In 1707, when Abraham Darby transferred his activities from Bristol to Coalbrookdale, many of his workers went with him; and Huguenot names in Cheshire today suggest that Spitalfields weavers were similarly moved to the north. But not infrequently the migration of a firm arose from a desire of the employer to free himself from irritating restrictions imposed by the craft traditions of the older towns: he must often have had a preference for new labour, and it was probably only the more skilled of his workers who were directly transferred.

There were obvious obstacles to migration over long distances. To the southerner Lancashire must have seemed almost as uninviting as Newfoundland and its denizens scarcely less uncouth than the Highlanders of Scotland. There were marked differences in dress between south and north, and it could not have been easy to change from a diet of wheaten bread, beef, and beer, to one of oatmeal, bacon, potatoes, and buttermilk. The northerners conversed in rough dialects that varied bewilderingly between places only a few miles apart. When the young Sarah Stubs of Warrington paid a visit to Sheffield in 1794 she wrote home to say she could hardly

[1] *Reflections on various subjects*, p. 41.

understand the people she met, in relatively polite circles, there: 'they talk so queer'.[1] A journey of fifty, or even twenty miles, must often have been a hazardous enterprise. No doubt bad roads and highwaymen mattered more to the well-to-do travelling by coach than to the working man going on foot; but other (and perhaps more serious) perils beset the way of the poor. In times of war there was the menace of the recruiting squad and the press-gang; and it was not uncommon for villagers to turn out and stone a wayfarer, for no better reason than that he was a 'foreigner'.

Under a system in which public relief was administered locally, well-to-do residents were usually concerned to keep down the number of the poor in the parish.[2] The Act of Settlement of 1662 was designed to check the influx of paupers to places where there were 'the best stock, the largest commons and wastes to build cottages, and the most woods for them to burn and destroy': its declared object was to keep people and resources apart. Under it the local authorities were given power to remove from the parish, within forty days, any entrant who might seem likely to become chargeable; and it is obvious that the knowledge that such power existed must have deterred men and women from moving.

There is reason, however, to think that the effects of the Poor Law on migration have been exaggerated. The statement of Adam Smith that 'it was often more difficult for a poor man to pass the artificial boundaries of a parish than the arm of the sea or a ridge of high mountains' may be accepted only if 'often' is interpreted as 'sometimes' or 'occasionally'. Both before and after a permissive Act of 1697, the churchwardens and overseers sometimes supplied a would-be migrant with a certificate, under which they accepted responsibility for his relief, or removal, if he should fall into destitution in another parish (which was usually specifically named). Single men who had migrated were rarely ejected from a parish unless they actually became chargeable; and where the incentive to move was considerable it is unlikely that considerations of settlement were allowed to stand in the way. 'We daily see manufacturers leaving

---

[1] T. S. Ashton, *An Eighteenth-Century Industrialist*, p. 142.

[2] Landowners sometimes pulled down cottages. Poor children were put out as apprentices in other areas. Employers gave preference to labourers whose homes were outside the boundaries, so that, when work was short, the burden of relief would fall on another parish. And contracts for hire were often limited to a period short of the full year that would have conferred on the worker a settlement in the parish in which he worked. Eleven and a half months became a very common term of employment.

The People of England 15

the places where wages are low and removing to others where they can get more money', it was said[1] in 1752. Agricultural labourers have never been the most mobile of workers, but in 1788, according to John Howlett, they 'ranged from parish to parish and from county to county unthinking of and undeterred by the law of settlement'.

Nevertheless, it seems probable that the existence of the Poor Law had an effect on the character of migration: it reinforced other tendencies that confined movement to relatively short distances. For, until 1795, the cost of returning a pauper rested, not on the parish demanding the removal, but on the parish of settlement; and in these circumstances overseers were reluctant to issue certificates except for places not more than ten or fifteen miles away. But if migration was only over short distances, how was the redistribution of population between south and north effected? The answer is almost certainly to be found in the process which Professor Redford has traced in the nineteenth century.[2] Areas of rapid industrial development (Lancashire, the West Riding, and Staffordshire, for example) drew in labour from neighbouring places. The gap created in these was filled by an influx of people from still farther afield, and so on. And thus a series of short waves was similar to a single long wave in effecting a northward and westerly drift. It should be added, however, that these statements are conjectural: figures are few, and in the absence of a periodic census it is impossible to trace in detail the caterpillar-like progression of labour across the face of the country.

Closely connected with the movement from region to region was that from the rural to the urban area. In the towns, burials usually exceeded baptisms. Yet the towns grew. They must have refreshed themselves with people from the countryside; and, since the belief that large numbers were driven off the land is no longer tenable, the migrants must have been attracted to the urban centres. It was common for country lads to go to the town to learn their craft: 'a village', said Josiah Tucker, 'is never the seat of the Arts.' Of boys apprenticed in Sheffield 15 per cent. in the first half of the century, and 25 per cent. in the second half, came from outside.[3] But in

---

[1] *Reflections on various subjects*, p. 9.
[2] A. Redford, *Labour Migration in England, 1800-1850*, pp. 160-1.
[3] E. J. Buckatzsch, 'Places of Origin of a group of migrants into Sheffield, 1624-1799', *Ec. Hist. Rev.*, Sec. Ser., vol. II, No. 3 (1950), p. 305.

the newer towns it was often unnecessary to have served an apprenticeship in order to practise a trade, and men of native wit, though of little technical education, found in places like Birmingham, Manchester, and Leeds a climate favourable to their enterprise. Above all, wages were higher in the towns: Arthur Young observed a progressive rise as he moved from the deep country towards London, and later investigators have recorded the same inverse variation of wages with distance from other large towns. Obstructions imposed by the Poor Law had little effect, it would seem, in checking the urban migration. For, though in the small rural parishes no intruder could hope to escape the notice of the overseer, in the large town it was clearly impossible for the authorities to keep an eye on every newcomer. The rise of the eighteenth-century town is a subject that would well repay investigation. All that can be said here is that it came about by an absorption of people from outside, and that the migration was generally over relatively short distances. Between 1700 and 1749 of the immigrant apprentices to Sheffield only about 22 per cent. came from places more than twenty miles away. And though, as time went on, the area of recruitment widened, between 1750 and 1799 the percentage was still slightly below 30.[1] Stress should be laid on the fact that it was not the old ecclesiastical or administrative centres, but new, unincorporated towns that saw the most rapid growth. The point was well made in the following passage:

'Corporation laws are trifling restraints in appearance, yet trifling restraints retard the growth of cities: and that so effectually and certainly that there is not a single city in England at this day on the increase; whilst most of our free towns, tho' with manifest local disadvantages get all the trade from them, and daily advance in wealth and numbers.'[2]

The eighteenth century thus saw important changes in the distribution of the English people. The counties in which agriculture predominated and the old textile regions lost ground to the northwest and the west-midlands, where manufactures of various kinds were expanding. In 1700 Lancashire was already well peopled: according to estimates which may be faulty but cannot be greatly in error, it came sixth, among the counties, in order of density. But in 1801, with 353 people to the square mile, it was far more thickly populated than any other region outside the metropolitan area. Over

---

[1] *Ibid.*, p. 304.   [2] *Reflections on various subjects*, p. 87.

the country as a whole the range of the figures had widened: people were less equally spread than at the beginning of the century. Most men still spent their lives in a rural setting. But many hamlets and villages had become industrial towns. And, as with the growth of numbers, so with the concentration of people, the chief developments came about after 1750.

II

Economic historians, like other students of society, are concerned primarily with groups. Their subject is not Adam, a gardener, but the cultivators of the soil as a class; not Tubal-cain, a skilled artificer in brass and iron, but metal workers or industrialists in general. They deal less with the individual than with the type. Some, indeed, write their stories in terms of *the* capitalist, *the* entrepreneur, *the* wage-earner, as though for each group there were an ascertainable norm, divergences from which might safely be ignored. In fact, the samples available to historians are rarely, in the statistical sense, random, and, in any case, are usually too few in number to support large generalisations. Any statements made about English society in the eighteenth century must be regarded as tentative. Men and women were highly individual: the one characteristic they had in common was a refusal to conform.

Social structure and human relations varied from one part of the country to another. In large areas of the midlands men still worked the land by methods, and under tenures, similar to those of Tudor, or earlier, times: in other parts of the country conditions were not unlike those of our own day. The tin miners of Cornwall, the lead miners of Derbyshire, the coal miners and iron smelters of the Forest of Dean, and the cutlers of Sheffield were all subject to corporate regulation, much as their forebears had been. But the coal miners of Northumberland and Lancashire, the copper miners of Anglesey, the artificers of Birmingham and the Black Country, and textile workers in all parts of the country knew little of such jurisdiction. Great corporations, including the East India Company, controlled large sections of the overseas trade based on London, and there were still societies of Merchant Adventurers at Bristol, Newcastle, and Hull. But at other outports, like Liverpool, no such ancestral form of organisation is to be found. In the textile areas of Lancashire, Yorkshire, and the West Country, the unit of production was the individual who worked in his own way and determined his own hours of labour: in the iron industry of Shropshire, Staffordshire, and South Wales it was a group of specialised wage-earners working under close supervision for a fixed daily number of hours. Men lived their lives in a provincial or parochial setting. The territorial aristocrat wielded more power over his neighbours than the sovereign state operating from London; the justice of the peace, on the spot, mattered far more than the civil servant in the metropolis. The market town where local produce was sold, local incomes spent, and local prices determined, was of more account than Smithfield or Mark Lane. If men wanted to improve a road or a river, construct a harbour, build a hospital, or initiate a new business enterprise, it was on local, rather than central, resources that they drew. Loyalties attached to the country house, the county regiment, and the regional hunt; affections were centred on the parish church, the village green, and the local inn.

Another feature of eighteenth-century life was the strength of the ties of kinship. The landed estate belonged less to the individual than to the family: it was the source of income not only of the man who, by law or custom, was termed the owner, but, through jointures and portions, of the women related to him by blood or marriage. Entails, made between father and son, were commonly renewed, generation after generation; and hence continuity of the union between the family and the land was assured. In the Church, in Parliament, and in the Law Courts, heredity counted for as much as zeal or eloquence. In agriculture the most common (or at least most commonly approved) unit was the farm that could be worked by a man and his sons.[1] The merchant house consisted generally of a small group of near relatives: the head of the family might be in charge of the business in London, while his brothers or sons acted as representatives in foreign parts. The Bubble Act of 1720 prohibited joint-stock companies, except under charters not readily granted; but, even if there had been no legal obstacle, it is doubtful whether industrial concerns would have drawn much of their enterprise from sources outside the family circle. Ironmasters wedded the daughters of ironmasters, potters the daughters of potters; and hence arose dynasties (such as those of the Darbys and the Wedgwoods) as powerful in industry as royal houses in international

---

[1] Andrew Wedderburn, *Essay upon the Question, What proportion of the produce of arable land ought to be paid as rent to the landlord?* (1766), p. 6.

politics. Some trades were open only to the sons of tradesmen. Weavers, stockingers, nailors, and many other of the 'poor and industrious' were assisted in their work by their wives and children; and even in the factories wages were usually paid not to the individual woman or child, but to the husband as head of the family.

Second only to kinship was religious persuasion. That there was a close connection between theology and economic progress was almost taken for granted by contemporary writers. 'Even religious duties themselves', wrote a publicist, 'may receive an higher lustre, and grow more divine by conspiring to promote the service of man on earth.'[1] But not all creeds lent themselves fully to this happy conspiracy. 'It is a point frequently urged by politicians and divines', the writer continues, 'that the Protestant religion is better calculated for trade than the Catholic; and the same have objected to the Methodists that theirs is not a religion for a trading people.'[2] Whatever may be thought of the apologetics, there can be little doubt that the precepts of the Catholic Church had not been framed with an eye to the development of manufacture and trade. Catholic teaching on usury, though less rigid than is sometimes supposed, put barriers in the way of dealings in money and speculative enterprises in general. The maintenance of religious houses and the observance of holy days hampered the expansion of output in France, and intolerance of heretics drove from the country a large part of the more skilled, ingenious, and hard-working of the subjects of Louis XIV. It is possible to cite instances of Catholics, like the Wrights of London and John Curwen of Cumberland, who made their mark on English finance and industry; but such men were few. Most Englishmen adhered to the established Church: many, perhaps most, of the leaders of industry and commerce were episcopalians. Foreign observers, like Voltaire,[3] were, however, impressed

---

[1] *Reflections on various subjects*, p. 34.

[2] *Ibid.*, p. 35. The writer goes on to remark that 'The great and religious Mr. Boyle gives it as one argument for propagating the Gospel in foreign parts, that if the converts could but learn so much of Christianity as to go cloathed, it would greatly add to the sale of our manufactures.' The benefit of the association, it was held, was mutual. 'As Religion has thus displayed her goodwill to man by condescending to some commercial ends, so Commerce has, on her part assisted Religion. Many of our modern churches owe their being and foundation to taxes on trade, not less than that grand mosque built by Solyman the First, from an imposition on all Christian commodities, which gave occasion to an observation that the Sultan was resolved to go to heaven, but unwilling to put the Turks to any expense in his journey.'

[3] Voltaire, *Letters on England, passim.*

by the part, wholly disproportionate to their numbers, played by dissenters in the economic life of the country. Many attempts have been made to explain this in terms of nonconformist doctrine or discipline, but none of these carries conviction. For the men who were responsible for the technical and economic innovations of the century derived their faiths and manners of life not only from Calvin but also from Luther, not only from George Fox but also (in spite of the opinion cited above) from John Wesley. Some held to the doctrine of election, others allowed scope for the exercise of free will. Some sought to walk by the guidance of the Authorised Version, others by that of the inner light. Some held to justification by faith, others stressed the efficacy of works. Some were 'quietists', others 'enthusiasts'; some rational, others mystical, in tendency. Some belonged to highly organised, hierarchical sects, others to independent, self-governing congregations. Among such diversity it is not easy to detect any common element other than that of dissent, and this alone would hardly seem to offer a clear promise of worldly success. The opinion has been expressed elsewhere that it was in the education afforded by the nonconformist bodies that the key to the riddle is to be found.[1] It was in the chapels and meeting-houses that minds were quickened and purposes shaped, and in the dissenting academies that the children of the middle classes acquired the habits of thought, the knowledge of languages, the familiarity with elementary science and book-keeping, that were often the foundation of material prosperity.[2] When to deeply held conviction there was added trained intelligence, and when these were associated with strong family sentiment and attachment to the local community, the moral force generated was intense. No crude theories of the forceful, self-seeking capitalist, ruthlessly exploiting his fellows, can explain the achievements, generation after generation, of the Darbys of Coalbrookdale, the Gurneys of Norwich, the Walkers of Rotherham, the Gregs of Styal, and many other Quaker, Methodist, and Unitarian families.

English society was a class structure. But class never hardened into caste. In France, the line between the nobility and the rest was sharply drawn, and in other countries of Europe, including Russia, there were few, if any, social groups interposed between landlords and peasants. In England, the aristocrats merged with the squires,

[1] T. S. Ashton, *The Industrial Revolution*, p. 19.
[2] H. McLachlan, *English Education under the Test Acts, passim.*

the squires with the freeholders and farmers. There was no sharp distinction between those who derived their incomes from rents and those who lived on the gains of commerce or industry. Landed families, like the Egertons of Tatton, put their younger sons into trade; and great proprietors, such as Lord Middleton, the Duke of Bridgewater, and Lord Gower, busied themselves with transport and manufacture. Successful traders, manufacturers, and bankers usually sought, before they ended their days, to become country squires. They might hope to set their sons among the great of the land, or at least to marry their daughters in such a way as to ensure that their grandchildren would be ennobled. Gregory King distinguished the merchants and traders by sea from the merchants and traders by land. No doubt the first were generally more wealthy and more highly esteemed than the second, but the difference can never have been clear-cut. Small-scale dealers, like William Stout of Lancaster at the beginning of the century, and relatively small-scale manufacturers, like Peter Stubs of Warrington, at the end of it, frequently ventured into overseas commerce. The very word 'merchant' was ambiguous. It was used to cover not only the wholesale trader in distant markets, but also the stock jobber, loan contractor, banker, exchange broker, and bullion dealer of London. Under conditions of industry in which an employer not only gave out work but also bought raw material and marketed the product, it is not surprising that many manufacturers assumed the title (as they had assumed the function) of merchant. Nor was the word 'manufacturer' more precise: it related, on the one hand, to the man who employed several thousands of his fellows, and, on the other, to the small craftsman who worked at the loom or the bench aided by a single apprentice or journeyman. Even among what were described as the poor there were many gradations, ranging from ingenious millwrights, like William Murdoch, through the skilled artisans, the servants in husbandry, the common labourers, and women spinners, to 'petty craftsmen, pretending to trades merely ostensible by which it is impossible to make a livelihood',[1] and to the paupers who had, and often sought, no other designation. These class gradations were rungs of a ladder on which many climbed and descended. Their existence, and the differences in incomes with which they were associated, may be

---

[1] William Godschall, *A General Plan of Parochial and Provincial Police* (1787), p. 5. He includes in this category 'seive and chair-bottomers, basket-makers, and vendors of matches, laces and garters'.

obnoxious to modern sentiment, but they led to emulation; and the fact that the rungs were many and closely spaced made it easy for ability to be equated to task. They were the product of centuries of history—a fact that has not been sufficiently appreciated by those who, looking at English progress in technology and wealth, lightly assume that similar results can be obtained with equal speed (and less social disharmony) in communities of undifferentiated peasants today.

The table in which Gregory King set forth his 'scheme of the income and expence of the several families of England' in 1688 cannot be statistically accurate.[1] But it merits attention as the work of a close observer of English society. It arranged the families (each consisting not only of parents and children, but of dependents, including servants and retainers) according to 'ranks, degrees, titles and qualifications'; and, as might be expected, the order conforms closely with that of incomes.[2] Temporal lords, with an average income of £2,800, head the list: cottagers and paupers, with an average of £6 10s., come almost at the end of it. If King's guess was right, about a quarter of the national income went to 3½ per cent. of the families, and a half to 18 per cent. of them. Three-fifths of the families of England shared among them little more than a sixth of the total income of the country. Unfortunately, it is not possible to give comparable estimates for the eighteenth century: it is probable, however, that the growth of manufacture brought an increased proportion of people into the middle ranges. What is certain is that the differences of income remained great: eighteenth-century society was emphatically not egalitarian.

A feature of the table is the division of the people into two approximately equal groups: those increasing the wealth of the country and those decreasing it. Among those increasing wealth are the lords, baronets, knights, esquires, and gentlemen, as well as the merchants, freeholders, farmers, shopkeepers, and artisans: those decreasing it are not only the vagrants and paupers, but also labouring people, out-servants, cottagers, and common seamen and soldiers. The criterion is whether income exceeded or fell short of expenditure, whether people saved, or were liable to have resort to the charity

---

[1] Printed in G. Chalmers, *op. cit.*, pp. 424-5.

[2] The chief exception relates to the clergy. Though the twenty-six spiritual lords had, according to King, an average income of £1,300, there were 2,000 clergymen with an average of £60, and 8,000 with £45 a year. The first group were roughly as well off as naval and military officers.

of others or to poor-law relief. The distinction made by King points to the existence of a mass of people who could never be sure of full employment. One of the greatest preoccupations of thoughtful men in this age was how to set the 'poor' to work, and to see that they laboured for the greater part of their time. It was realised that the maintenance of employment required a continuous flow of capital: hence the stress laid by a long line of publicists on the virtue of saving.

Individually, the largest savers were, according to King, the nobility and gentry and the merchants trading overseas. It is sometimes suggested that the English landed proprietors of the eighteenth century were spendthrift by nature. It is true that their manner of living does not suggest parsimony, and instances of aristocrats who wasted their substance at Court or the gaming table are not hard to come by. But the class as a whole did not dissipate its patrimony. Landlords vied with each other in the splendour of their houses and stables, the breadth of their estates, the breed and weight of their cattle, and the yield of their acres. If they spent lavishly it was mainly on durable goods (including paintings and statuary). They sought prestige (as Jacob Viner has said) in conspicuous investment, rather than in conspicuous consumption. And, though their chief object was to build up their family fortunes, on balance their saving was beneficial to the public. That the merchants should have been noted for thrift need cause no surprise: they required capital for their own ventures and, as has already been said, their ambition was generally to accumulate wealth sufficient to found a family, buy an estate, and rise in rank and esteem. But though, individually, landlords and merchants were the greatest savers, their numbers were relatively small. If King is to be relied on, the largest contribution to resources came from the thrifty freeholders and farmers who, in 1688, made up the bulk of the middle classes. Lawyers, dealers, officials, shopkeepers, and tradesmen also added to the fund of savings, and even the artisans and craftsmen contributed their mites.

Nothing impresses the student of eighteenth-century estate and business records more than the care with which resources were administered, and the attention paid to small items of profit or loss. The habit of calculation, natural enough in the market place, is exhibited even in the sphere of domestic and family relations. Women were as meticulous in recording details of household expenditure as their husbands were in keeping their business accounts. Under a will of 1765, John Holt of the Warrington Academy, divided his substance among his relatives in strict proportion to the varying degrees of kinship, the shares ranging from one 'part' to one sixty-fourth of this. The same precision is exhibited in the relations between the merchant John Pinney and his children. Having acquired in 1743 a life interest in an estate entailed on his sons, he took care, on the birth of each male child, to open an account in which the infant was debited with the cost of the midwife, the fee for the baptism, and its share of the wages of the nurse.[1]

But, if many Englishmen were cautious and parsimonious, others were carefree and adventurous with their money. London was a gambler's paradise. The man about town could indulge his taste for cards (ace of hearts, faro, basset, or hazard) at fashionable clubs like White's, the Cocoa Tree, and Almack's. He might bet at the cockpit in Goodman's Fields, or lay wagers on the swiftness of his footman in Hyde Park. He could win or lose on the fortunes of wrestlers or boxers, and back horses at Newmarket, Ascot, or Epsom. Ladies of rank not only patronised, but sometimes themselves set up, gaming houses; and opportunities of sessions at the card table were among the attractions of the season at Bath. The lower orders would lay money on their agility in field sports, or their dealings with skittles and dominoes in the tavern. All classes betted on the turn of a coin, the throw of the dice, the result of an election, the prospects of the demise of a celebrity, or the sex of an unborn child. Gambling in church was not unknown, and, on one occasion at least, a wager was made and settled on the floor of the House of Commons.[2]

The widespread propensity to gamble was a matter of concern to statesmen, especially as it affected the industrious poor. But the state itself fed the flames. Between 1694 and 1784 it set on foot no fewer than forty-two lotteries, each associated with a loan or a conversion of debt.[3] And though governments sought to preserve to themselves this method of raising money (private lotteries were made illegal in 1721) the public paid more heed to the example than to the precepts of its rulers. Occasionally special Acts were passed to

---

[1] R. Pares, *A West-India Fortune*, p. 68.

[2] It was between Pulteney and Walpole and arose from a misquotation of Horace. When the Speaker gave the verdict against him, Walpole flung a guinea to the winner with the comment that it was the first public money Pulteney had touched for a long time. John Ashton, *The History of Gambling in England*, p. 156.

[3] Jacob Cohen, 'The Element of Lottery in British Government Bonds, 1694-1919,' *Economica*, new ser., vol. XX, No. 79.

allow favoured individuals or semi-public authorities to make use of the device. In 1738 a body of commissioners was allowed to appropriate 15 per cent. of the receipts from the sale of lottery tickets for the building of Westminster Bridge. In 1753 the proceeds of another lottery went to the purchase of the collection of Sir Hans Sloane, and so to the founding of the British Museum. In the slump of 1773 the Adam brothers, embarrassed with obligations arising from the building of the Adelphi and from their holdings in the Carron Iron Company, were allowed to offer part of their properties as prizes in a lottery. And in 1784 Sir Ashton Lever of Alkrington was permitted to dispose of his collection of fossils and shells by the same means.[1] In every lottery there was an interval between the sale and the drawing of tickets; and, since during this period the government enjoyed an interest-free loan, it had every reason to see that it was a long one. For several months tickets changed hands as rapidly as guineas; and shopkeepers, and even barbers, attracted custom by offering their clients the chance of a fractional share in a ticket. Licensed dealers (including one with the seductive name of Richards, Goodluck & Co.) reaped large gains, and, during the days of drawing, speculation amounted to frenzy.

Insurance, which is usually thought of as a method of reducing risks, was often used as a means of gambling. In the curious form of the tontine it offered security combined with large benefits to the fortunate survivors. Other devices were less free from objection. It was common for people to buy annuities on the lives of selected, healthy, male children—the dangers incidental to child-bearing excluded females—or to take out policies on the lives of both children and adults who were thought to be nearing their end. For a small premium it was possible to obtain a fairly large sum if, on a given day, a lottery ticket of a specified number were drawn from the wheel, even if only as a blank. And reinsurance, a method of spreading risks, was for long employed as a cover for wager policies.[2]

Gambling extended into the field of trade and finance. Englishmen in the West Indies betted on the size of the crops of plantations[3] (just as the Dutch betted on the date of the first catch of herrings) and Englishmen at home laid wagers on the yield of the hop duties.[4]

---

[1] John Ashton, *A History of English Lotteries, passim.*
[2] B.P.P., *Report of Commrs. of Inland Revenue, 1856-67.* 1870, vol. i, p. 79.
[3] R. Pares, *op. cit.,* p. 76.
[4] H. H. Parker, *The Hop Industry,* p. 45.

C

There was much speculative dealing in the various forms of the public debt and in the stock of the chartered companies: the South Sea affair of 1720 was neither the first nor the last of a series of Stock Exchange booms. Time dealings were common throughout the century: Sir John Barnard's Act of 1733, which sought to suppress 'puts' and 'calls', was a dead letter. Merchants bought and sold foreign bills not only to get remittance or payment in respect of goods, but also to take advantage of variations in exchange rates between one centre and another, to anticipate appreciations or depreciations of currencies, and to obtain present command over money in return for a promise of future payment.[1]

There is a danger of laying stress on extremes. It was neither the misers nor the gamblers that built up enduring concerns in industry and trade. Men like Abraham Darby, Richard Arkwright, and Robert Peel administered their incomes with care but were willing to bear reasonable risks. It is often said that these and other pioneers set up their concerns with their own savings and expanded them with capital spun out of the businesses themselves. It is true that self-financing was a marked feature of the period, but it would be an error to consider it as universal or to think of the markets for capital as circumscribed. The East India Company, the Bank of England, the South Sea Company, and a number of insurance societies were able to obtain resources by the sale of their stock; and, since 1694, when the national debt came into being, the government itself had raised funds not so much by personal appeals to individual merchants as by disposing of annuities, exchequer bills, navy bills, and other transferable securities to all comers. Funds for investment were more abundant in London than in the provinces; and London merchants not only financed country manufacturers with whom they traded, but also supplied capital for public works at Liverpool and were heavily involved in mining and iron smelting in South Wales.[2]

At Bristol, Hull, and Newcastle, loanable funds were also relatively plentiful, and from each of these a stream of resources flowed to the industrial areas. Rates of interest were generally lower in the metropolis than in the outports, and lower in these than in the manufacturing towns and villages; but the channels of finance

[1] Dealings in exchange could be used as a means of evading the usury laws.
[2] A. P. Wadsworth and J. de L. Mann, *The Cotton Trade and Industry of Lancashire 1660-1780*, pp. 215-17; A. H. John, *The Industrial Development of South Wales*, ch. II.

were sufficiently deep and extensive to ensure that changes at the centre were followed by similar changes throughout most of England.

In a century of costly wars, the chief borrower was the government. When its need for resources was urgent it offered a high rate for money, and the prices of stocks bearing a fixed rate of interest accordingly fell. The yield on government stocks gave the ply to the whole system of interest rates. Most of the borrowing and lending between individuals was on mortgages or bonds. The eighteenth-century mortgage differed from its modern counterpart in that it was rarely a contract to pay a fixed rate of interest for a defined number of years: either party might withdraw on giving six months' notice, or might insist on a change of interest as a condition of continuing the loan. Moreover, a mortgage might change hands by assignment or sale: it was a flexible form of security. The rate it bore (and the same was true of that on bonds) tended to follow, with a lag of six months or a little more, the rate yielded by government securities.

There are some industries to which interest is of special importance as an element in costs: they are those that require a large initial outlay, which can be recovered only by returns extending over a long series of years. In the eighteenth century, as at other times, a rise or fall of one per cent. in the price of capital might affect the decision as to whether to build a house or a ship, sink a mine, or construct a canal or a turnpike road. Activity in such undertakings varied closely with the price (or inversely with the yield) of government stock. In most industries, however, a rise or fall of one, or even two, per cent. in the rate paid on borrowed money made little difference to the ratio between costs and returns. Nevertheless these industries also varied in activity with the price of the Funds. For this price was an index not only of the needs of the state for money, but also of the supply of savings available for investment. In a free economy the rate of interest would have been the instrument by which the distribution of loanable funds was determined: the state and its subjects would have competed on equal terms for whatever supply was forthcoming. In fact, the rationing of funds was not effected automatically in this way. For the usury laws set a limit (6 per cent. before 1714 and 5 per cent. thereafter) to the rate any member of the public might bid. These laws were generally defended on the grounds of equity: the case for them rested on the supposed need to protect the borrower from extortion. But their effect was, at times, to divert resources from landlords, farmers, merchants, and manufacturers to the state. However great the need of such men might be, however rich the returns their projects might have yielded, they were not allowed to offer more than 5 per cent. for money. The government, on the other hand, might offer whatever rate it pleased on a new loan, and might suffer the price of an old loan to fall to such a level as to make its yield to the purchaser exceed that obtainable elsewhere.

In the following pages frequent mention will be made of *the* rate of interest. The reference is to the yield on government stocks. Other rates—those on mortgages, bonds, and bills—were generally higher, for there was usually a greater element of risk in lending to private individuals than in entrusting money to the state. But all rates rose and fell together within the upper limit set by the law. The reasons for selecting the return on government annuities as the index of the supply of money are many. Though large numbers of quotations of the interest paid on mortgages are available,[1] it is not possible, as it is with the yield on the Funds, to observe day-to-day changes. Though the interest on mortgages and bonds was elastic, it moved up and down by stages of one, (or, at the best, of a half of one) per cent.: yields on government stock were more sensitive. But the chief reason is implicit in what has been said above. Commercial rates are an index of the degree of scarcity of capital only within limits. If the yield on government stock rose as high as 4 or $4\frac{1}{2}$ per cent. that on mortgages, bonds, and bills might stand at 5 per cent. Beyond that it might not legally go; and though evasion was by no means unknown, the penalties were high and the law was generally respected. There was no limit to the yield on government stock: it is an accurate index of the supply of loanable funds at all levels.

This disquisition may seem somewhat theoretical and remote from reality. Had the small trader or manufacturer in the provinces any reason to take account of the financial operations of his rulers in London? A single example may suffice to show that he had. In 1757 Jedediah Strutt and his brother-in-law, William Woollatt, were seeking a patent for their Derby-rib stocking frame, and were in need of money to exploit the invention. Strutt's wife, Elizabeth, had previously been in service with the Rev. Dr. Benson who was, at this

[1] Mr. D. Joslin has assembled a long table of the rates at which the London banks lent on mortgage.

time, living in London. In the hope of inducing him to finance the new enterprise, she travelled from Derby to his home in Goodman's Fields; but in a letter dated '3 May 1757 6 o'clock in the morning' she had to inform her husband that her mission had failed:[1]

'Ye Dr. is pritty well again and I have acquainted him with our scheme which, so far as he understands it may do very well and he will do all he can for us, and wou^d willingly supply us with ye money. Mr. William Cook at ye same time wanted to Borrow of him one thousand pound (and) in order to furnish boath of us he went to ye Bank to sell out but ye War makes ye Stocks run so very low y^t he will loose a Hundred pound if he sells out now, and they will rise as much in proportion if there comes a peace so y^t he would rather chuse to Borrow for his own use than loose so much money. Y^s is a great Disapointm^t to me and yet I cannot desire him to act otherwise. He says y^t he or anny Body will be as much obliged to you for 4 pr cent for their money as you will be to them for it.'

In May 1757 the 3 per cent. annuities stood at 90. Dr. Benson was unwilling to forgo the chance of a recovery in the capital value of his holding. But he may also have reflected that the yield on investment in the annuities was now 3·3 per cent. and that the difference between this and the 4 per cent. offered by Strutt was small compensation for the additional risk.

On the whole, eighteenth-century England was blessed with cheap money. Between 1726 and 1776 there was only one brief occasion (June and July, 1759) when the return on the Old annuities exceeded 4 per cent. Contemporaries were aware of their good fortune, and many of them, indeed, overstated the benefits that might come from still lower rates of interest. But from time to time landlords, farmers, and manufacturers were unable to obtain the resources they needed, not because savings had dried up, but because these had been diverted to the state. Unable to raise money by the offer of higher rates of interest, men were forced to sell their assets for whatever these would fetch. The violence of the fluctuations by which the century was marked is to be attributed less to the speculative tendencies of producers and traders than to the operation of the laws against usury.

[1] R. S. Fitton and A. P. Wadsworth, *The Strutts and the Arkwrights, 1758-1830* (1958), p. 30.

CHAPTER TWO

Agriculture and its Products

I

ACCORDING to some accounts, English agriculture in the eighteenth century was a way of life to which every right-minded townsman must look with nostalgia. According to others, it appears as little more than a battlefield on which whole classes were defeated by an hereditary enemy and virtually extinguished. It is necessary to insist that it was, above all, an industry. It employed more resources than any other. Those who controlled it were no less concerned than iron masters or cotton spinners to maximise their incomes and properties. They were no more, and no less, public spirited than their fellows. The idea (held by some modern as well as contemporary writers[1]) that they rendered a service greater than that of other industrialists, is as false as the idea that they rendered no service at all. Agriculture had its peculiar features. Its techniques differed from place to place. Of its varied products a large part was consumed on the spot. The esteem that attached to ownership of the soil affected its progress. But generally, like other callings, it was ruled by the forces of the market.

Most of the holdings were mixed farms. Grass and grain provided food for livestock, and the livestock enriched the meadows and cornfields: there were few rural areas that were unable to produce the bread, beer, meat, and milk they required. Each part of the country, however, specialised in the things which, by situation, soil, or climate, it was best fitted to produce. Over the greater part of England the staple breadstuff was wheat or maslin (rye and wheat); and in Wales it was barley. Dr. Johnson's celebrated designation of oats as 'a grain, which in England is generally given to horses, but in Scotland supports the people', was only partially true, for the dietetic boundary was the Trent rather than the Tweed: in the six northernmost counties, at least, oatmeal rather than wheaten flour was the standard

[1] A distinguished modern scholar holds the view that 'it is the only industry of which it may be said that it produces wealth without destroying it'. C. S. Orwin in *Johnson's England* (edited by A. S. Turberville), p. 262.

Agriculture and its Products

food, and in Lancashire potatoes were becoming increasingly important.[1] Wheat, barley, oats, and rye were grown in most parts of England, but there was regional concentration on one or another of these crops. East Anglia and Essex, the south from Kent to Dorset, and most of the clay belt of the midlands were given up largely to wheat, as were also the counties of Monmouth, Hereford, and Pembroke. Barley was grown extensively in Cambridgeshire, Hertfordshire, Berkshire, Wiltshire, and parts of Yorkshire; and oats in Northumberland, Durham, Lancashire, Cheshire, Derbyshire, Nottinghamshire, and Lincolnshire. There was much inter-regional traffic: London drew its supplies from East Anglia and the valleys of the Thames and the Lea, Bristol from the areas about the Severn and the Wye, and the industrial area of Yorkshire from the East Riding and Lincolnshire.

Similar specialisation existed in the production of livestock. Land unfitted for other purposes was used to raise cattle and sheep, for, as Defoe observed, 'tho' the feeding of cattle indeed requires a rich soil, the breeding of them does not, the mountains and moors being as proper for that purpose as the richer lands'.[2] 'Black cattle' (i.e. store cattle) bred in Scotland were fed in Norfolk meadows: others raised on the mountains of Brecknock and Radnor were sent to the marshlands of Essex. But there were large areas in Leicestershire and Lincolnshire where the graziers bred and fattened their own cattle. Sheep were widely distributed. They were raised on the mountains of Wales and Cumberland, on the Pennines, the Cotswolds, and the Devonshire moors, on the downs from Sussex to Dorset, in Romney Marsh and the Fen country, but, most of all, in the belt of territory stretching between Warwickshire and Lincolnshire, and lying between the Great Ouse and the Trent. Horses were bred in Northamptonshire and Leicestershire, as well as in Durham, the North Riding, Suffolk, Devonshire, and Montgomeryshire. There was some specialisation of types: the best race horses came from Durham, the best coach or dray horses from Northampton, the best plough horses from Suffolk. Dairy farming was carried on extensively in Cheshire as well as in Wiltshire, Glamorgan, Huntingdon, and Suffolk (which, according to Defoe, was famous for the best butter

---

[1] Dr. Johnson's jibe drew from Lord Elibank the retort, 'Very true. And where will you find such men and such horses?' A. W. Read, 'A History of Dr. Johnson's Definition of Oats', *Agricultural History*, vol. 8, No. 3, p. 81.

[2] *A Tour through England and Wales* (Everyman ed.), vol. II, p. 59.

and perhaps the worst cheese in England[1]). There were orchards and hop fields in East Kent and Worcestershire, and most of the West country produced cider. Whey from the dairies, and mash from the cider presses, provided food for hogs, and hence bacon was a product of most of these counties. Geese and capons were raised in Sussex and Surrey, and turkeys, as well as geese, in Norfolk and Suffolk. Each year, in August, they set out in droves of a thousand or more for London, feeding on the stubble as they went, until the end of October 'when the roads begin to be too stiff and deep for their broad feet and short leggs to march in'.[2] About most large centres of population there were market gardens, fertilised by the refuse from the towns. And much of the land to the north and east of the metropolis consisted of meadows to provide fodder for its stables: haymakers often thronged the streets of London, seeking in good times diversion, in bad times relief of their distress.[3]

The impression left by Defoe of agriculture in the seventeen-twenties is one of enterprise, experiment, mobility. Turnips and clover had been introduced into the crop cycle in Norfolk. Stress is laid on fertilising the soil. On Salisbury Plain sheep are being folded on newly ploughed land, and thin soils, hitherto uncultivated, are producing good crops of wheat: 'if a farmer has a thousand sheep, and no fallows to fold them on, his neighbours will give him ten shillings a night for every thousand.'[4] At Cheddar where 'the whole village are cowkeepers, they take care to keep up the goodness of the soil by agreeing to lay on large quantities of dung for manuring'.[5] In Norfolk marl is being dug from underlying beds to mix with the light, sandy topsoil. Limestone is being carried across the mouth of the Thames, and so 'the barren soil of Kent, for such the chalky grounds are esteem'd, make the Essex lands rich and fruitful'.[6] Both Essex and Suffolk are making use of underground drainage. Experiments are being made in breeding: during the War of the Spanish Succession Durham turned 'from fine fleet horses for galloping and hunting, to a larger breed of charging horses, for the use of the general officers and colonels of horse. . .'.[7] All this was before what

---

[1] Ibid., vol. I, p. 53.   [2] Ibid., vol. I, p. 59.
[3] In June, 1766, 'the haymakers assembled at the Royal Exchange to the number of 440 persons, when a collection was made for them on account of the heavy rains, which prevented their getting work'. *Annual Register*, 1766. M. D. George, *op. cit.*, p. 357, refers to pitched battles between English and Irish haymakers at Kingsbury, Hendon, and Edgware in 1774.
[4] Ibid., vol. I, p. 285.   [5] Ibid., vol. I, p. 277.
[6] Ibid., vol. I, p. 99.   [7] Ibid., vol. II, p. 224.

Agriculture and its Products

is usually regarded as the age of improvement—long before the days of Coke and Bakewell. Defoe had an eye for whatever was striking or unusual, and sometimes he ran to hyperbole; but it is impossible to ignore the picture he presents. It is certainly not one of a community of supine peasants.

Systems of cultivation and tenure varied from place to place. Whether the fields were open or enclosed seems to have depended on configuration, soil and subsoil, the nature of the product, and the distance from centres of consumption.[1] The open-field system was unsuited to mountainous country or pastoral farming. There are signs that it had existed in some of the valleys of Devon and Cornwall, but none that it was in use there in the eighteenth century.[2] It was rarely found in Wales; it had gone, or almost gone, from the whole of the west country from Monmouth and Somerset to Shropshire; and it was hardly to be seen in Cumberland, Westmorland, and Lancashire. Though it might serve well enough the needs of areas in which men raised grain for their own consumption, it was less well adapted to those in which production was for sale. It was unknown in Kent, and the influence of the London market had led to its early decline in Middlesex, Essex, and a large part of Suffolk. Its province was the broad belt stretching from Hampshire and Dorset through the midlands to Yorkshire; though even in this there were many scattered areas of enclosure. In the early eighteenth century it would appear that about half the arable land of England was held in intermixed open-field strips.

The system was not without advantages. It has been said that the necessity of conforming to the communal time-table for sowing and harvesting, of following the agreed rotation, and observing the customs of the village or manor, ensured that cultivation did not fall below certain standards. Open-field farmers could agree to exchange strips so as to get larger contiguous holdings, and sometimes they added a crop to the customary cycle or made arrangements to hold meadows in severalty: there is no need to think of the arrangement as inflexible. But it was unfriendly to the individual who wanted to move ahead of his fellows. Rights of pasture on the stubble and

---

[1] For geological influences see, M. Aurousseau, 'Neglected Aspects of Enclosure Movements', *Economic History*, No. 2 (1927), pp. 280-3. It is pointed out that the early enclosures were on the fertile areas of the Tertiary formations and those of the eighteenth century on the stiffer lands underlaid by the Mesozoic rocks.

[2] R. R. Rawson, 'The Open Field in Flintshire, Devonshire, and Cornwall', *Ec. Hist. Rev.* (second ser.), vol. VI, No. 1, pp. 51-4.

common gave rise to interminable disputes. And the small size of the strips and furlongs, especially when these were separated by balks, made it difficult for any one man to drain his land properly. Even if his neighbours were willing to let him connect a surface ditch with one they had made themselves, there was little hope of the same indulgence for hollow, or underground drainage. The inability of the system to deal with the problem of wet lands, was perhaps the chief of its drawbacks.[1] It is significant that nearly all the improvements of agricultural technique of which there is record were made on land already enclosed or in process of enclosure.

II

The English countryside was studded with estates, each with its hall or manor house, garden, park, and circumjacent acres. These were most numerous in the home counties, but were also sprinkled about counties, like Durham and Lancashire, remote from the attractions of London and not remarkable for sport and rural diversions. They were usually spoken of as family seats. Some were owned by noble families like the Talbots and the Howards; others by families of lower degree (though not necessarily of shorter lineage) who formed the ill-defined class of gentry or squires; and yet others by men who had recently come to fortune by way of some trade or profession, and who (though no doubt they had had forebears and might have begotten heirs) could hardly be said to belong to families at all.

Throughout the eighteenth century the number of estates owned by members of the last of these groups increased. City men and manufacturers moved to the country, sometimes for the amenities it offered, sometimes as a step to Parliament, but often because they cherished the ambition of ending their days on at least level terms with the squires. At the same time noblemen were extending and consolidating their position on the soil. To ensure continuity of the estate, they created entails and nominated trustees whose terms of appointment, sympathetically interpreted in the law courts, enabled them to see to it that these were not broken. To ensure good administration they employed lawyers who were specialists in estate management, and engaged as their stewards men who had had practical experience of farming. And they ploughed back their profits into

[1] Stress is laid on this by G. E. Fussell, *The Farmers' Tools*, pp. 18-19.

the soil.¹ Some of them were explicit as to their motive: it was to regain the footing their families had lost in the political and social disorders of the seventeenth century. In a long series of letters to his son a northern landowner asserted roundly that merchants had usurped the position of power and esteem that rightly belonged to men of pedigree and title. The son was enjoined, first to seek a wife with £10,000 or £20,000 a year of her own, second to substitute leases at rack rent for the loose agreements that had hitherto been made with tenants, and thirdly to live unostentatiously and put each year several thousand pounds into a cumulative fund. If the son, in turn, were able to persuade his heir to follow the same course, a somewhat remote descendant of the writer might be able to occupy the position in society the essential title to which was birth.

As Professor Habakkuk has said, the estate was a unit of ownership, not of production. The owner looked to it to provide a stately house in gracious surroundings and to supply income in the form of rents and fines. He might have a home farm to supply the needs of his family and servants, but he rarely cultivated for the market: his function was to supply capital and general direction. The income from the estate was likely to be greater if the acreage were increased; if scattered portions of the land were exchanged for others so as to give a more compact unit of control; if the individual farms were of a size to utilise to the full the energies of the tenants; and if farmers and labourers were kept informed as to improvements in agricultural practice.²

The part played by the aristocracy in stimulating new methods of husbandry was not unimportant. Norfolk landowners, of whom those best remembered are Lord Lovell, Viscount Townshend, and Thomas Coke (afterwards Earl of Leicester), drained their wet lands, mixed marls with sandy soils, stimulated their tenants to grow root crops to provide fodder for cattle, and introduced more efficient rotations. Their methods spread, with modifications, to other parts of the country, and great houses (like Woburn, Petworth, and Wentworth) became centres of experiment and instruction on agricultural technique. But the story, and that of the parallel work of midland landowners in improving the breed of cattle and sheep, has been told so often that nothing more will be said of it here.

---

[1] H. J. Habakkuk, 'English Landownership, 1680-1740', *Ec. Hist. Rev.*, vol. X, pp. 1-17.
[2] *Ibid.*

Various steps were taken to extend the area of the estate and the control of the landowner over the soil. One was to buy out the rights of smaller squires and freeholders: in the first half of the century the numbers of both of these declined. Professor Habakkuk points out that squires with rentals of £800 or less were in a weak position at this time: unlike the great landlords and the new men from the city they had few, if any, outside sources of income; they were hit by the Land Tax; and they were unable to provide farms of the size, and with the equipment, that their larger competitors could offer. The small estate was no longer a viable unit.

Freehold is a term applicable to land rather than to cultivators. As historians of the seventeenth century have pointed out, some men who held agricultural land in fee simple were townspeople who let their holdings to tenants, and some who cultivated their freeholds also farmed other plots under leases. In various ways freehold shaded off into copyhold and other forms of tenure: there must have been many among the 180,000 so-called freeholders in Gregory King's estimate of 1688 who were not simple yeomen tilling their own acres and beholden to no landlord. But the yeomen existed, and there is a consensus of opinion that, however they might be defined, their numbers declined in the first half of the century. The idea that they were physically driven from the land is unsupported by evidence. Those who used to hold it laid stress on the martial qualities of this 'sturdy' class: if large numbers had been evicted they would hardly have gone quietly. But there are no records of agrarian risings or even local battles of any consequence at this time. The process was one of attrition. Landlords were exercising a demand for large-scale tenant farmers: they were reluctant to supplement the acres of the freeholders by letting to them land in small plots.[1] The man with a legal title to small portions of soil intermixed with those of the estate owner was an obstacle to consolidation and enclosure: it was politic to buy him out. The small freeholders were often willing to sell: they could not compete with tenants who had larger holdings and greater capital resources. Some of them may have left the country for the towns, but others may have used the money from the sale of their freeholds to buy stock and equipment and set up on a larger scale as leaseholders.

---

[1] H. J. Habakkuk, *loc. cit.* It is pointed out that the freeholder was likely to concentrate such improvements as he might make on the land he held in fee simple, and to neglect that he held under lease.

Agriculture and its Products

Conditions of tenancy were revised. From the point of view of the estate owner the traditional system of leases for lives had serious defects. The fine that the tenant had to pay on renewal might be substantial, but the annual rent was accordingly low. The impossibility of knowing when a lease for three lives would terminate was a handicap to good estate management. It probably deterred landlords from making capital improvements in the farm; and the tenant, with fines looming ahead, might put less into the soil, and more into cash balances, than he would otherwise have done. On the other hand, tenancies at will, or from year to year, were generally disliked by farmers, and their uncertainty told against efficient cultivation. Hence, though all forms of tenure existed side by side, it became increasingly common to let land for moderately short periods, such as seven, fourteen, or twenty-one years.[1]

It is impossible to obtain accurate statistical information about the course of rents. Each parcel of land was unique. Rents varied with the quality of the soil, access to water, proximity to markets, the state of the farm buildings, and the length of the lease. Generally the Land Tax was paid by the owner, and tithes and parochial rates by the tenant, but there were local variations of practice. Corporate bodies, like Eton College, sometimes varied their rents annually with the movement of prices of wheat.[2] But even where there was no such formal arrangement, landlords were accustomed to make abatements in years when the prices of grain were unduly low or when local harvests were poor. In the first half of the century the course of wheat prices was downward, and those who have made a close study of estate accounts consider that the trend of rents was in the same direction. The fall of both prices and rents was not unconnected with the substitution of leasehold for freehold.

Farms were growing larger. The old idea that a holding should be of a size sufficient to maintain a family and no more was beginning to give way to a belief in bigger units of production. These could introduce division of labour, utilise larger capital, and supply markets more efficiently.[3] The notion grew up that ideally the

---

[1] For the merits and demerits of different types of leases see W. Marshall, *The Landed Property of England* (1804), pp. 358-67.

[2] The tenant had the option of paying in money or kind. Alexander Wedderburn, *op. cit.*, p. 36.

[3] The case for the large farm is argued in *An Inquiry into the Prices of Provisions and the Size of Farms, with Remarks on Population. By a Farmer* (1773)—attributed to John Arbuthnot.

produce should be about three times as great as that consumed by the farmer: of the surplus half should cover the cost of seed, implements, and hired labour, and half should go as rent.[1] In the early part of the century many arable farms were of 200 acres or more, and in the grassland areas they were far larger: in Leicestershire, according to Defoe, it was not uncommon for graziers to take large tracts and pay rents of as much as £500.[2]

The conversion of freehold into leasehold was a step in a larger process of enclosure. The word was taken to include the consolidation of scattered holdings, the extinguishing of common rights in the arable fields and meadows (and of the appendant or appurtenant rights in the waste) as well as the fencing or hedging of each man's portion. A cultivator might consolidate by arrangement with his neighbours, but, if his land were subject to common grazing after the crop had been got in, he could not refuse entry to others. An estate owner might reassert rights that had tended to be disregarded, and he might attempt to enclose part of the waste to create or extend a park. But if he did so, he was likely to meet with resistance. Defoe describes how a Buckinghamshire squire, Mr. Guy, 'presuming upon his power, set up his pales, and took in a large parcel of open land, call'd Wiggington Common'; and how 'the cottagers and farmers oppos'd it, by their complaints a great while; but finding he went on with his work, and resolv'd to do it, they rose upon him, pull'd down his banks, and forced up his pales, and carried away the wood, or set it on a heap and burnt it; and this they did several times . . .'.[3] For enclosure in the full sense it was necessary to deal with the land of the village or parish as a whole. If an estate owner were able to buy up all the rights of his neighbours he could do as he pleased with the land. But this involved great expense. Sometimes it was possible to enclose by agreement (a word that must not be taken to exclude pressure and threats of suits in Chancery). But more often procedure was by Act of Parliament. A public meeting was called; notice of an intended petition was affixed to the Church door; and the draft of a private bill was prepared. The bill went to committee and, often after long debate, the Act emerged. It appointed commissioners, usually three in number, to execute its provisions, and

---

[1] The idea of the tripartite division of the produce runs through the pamphlets of the century. It is discussed, in particular, by Alexander Wedderburn, *op. cit.*, pp. 4-6.

[2] Defoe, *op. cit.*, vol. 2, p. 89.   [3] Defoe, *op. cit.*, vol. 2, p. 15.

Agriculture and its Products

these were assisted by surveyors and valuers. An estimate was made of the value of the land as it would be after enclosure, and the existing value, as ascertained from the rent roll and other sources, was deducted. The difference was available to cover legal and parliamentary expenses, the making of roads and drains, and any provision Parliament made for the relief of the poor who might suffer from the redistribution of property. The work of the commissioners was arduous, and sometimes it extended over ten or twelve years.[1]

At each stage there was a possibility of injustice; and it is beyond doubt that the scales were loaded against the smaller cultivators who often had no taste for enclosure. But the careful researches of W. E. Tate and others show that relatively few counter-petitions were put forward and that there was no systematic furthering of their own interests by members of Parliament.[2] According to Professor Gonner the work of the surveyors 'appears to have been honestly, if not always well, done, and to have been marked by a rough and ready fairness'.[3]

In the first half of the eighteenth century enclosure by Act was unusual. Where the advantages of enclosing were obviously great, as on the light soil of Norfolk or the dairy lands of Gloucestershire and Wiltshire, there was probably no need for legislation.[4] And in other places where the number of freeholders was small, it was usually possible to reach some form of agreement. But there was at this time no great pressure to enclose. After the Union of 1707 the number of Scots cattle driven across the border each year was so great, it was averred, as to bring depression to the English graziers. Farmers turned to the production of cereals, and for this reason, as well as for others connected with the currency, the trend of prices of grain-stuffs and animal products was downward. Most sessions saw two

---

[1] For a detailed account of procedure see E. C. K. Gonner, *Common Land and Inclosure*, pp. 71-95.

[2] W. E. Tate, 'Members of Parliament and the Proceedings upon Enclosure Bills', *Ec. Hist. Rev.*, vol. XII, Nos. 1 and 2, pp. 68-75; 'Opposition to Parliamentary Enclosure in Eighteenth-Century England', *Agricultural History*, vol. 19, pp. 137-42; 'Members of Parliament and their Personal Relations to Enclosure', *idem*, vol. 23, pp. 213-20; 'The Commons' Journals as Sources of Information concerning the Eighteenth-Century Enclosure Movement', *Economic Journal*, vol. LIV, pp. 75-95; M. W. Beresford, 'Commissioners of Enclosure', *Ec. Hist. Rev.*, vol. XVI, No. 2, pp. 130-40.

[3] E. C. K. Gonner, *op. cit.*, p. 76.

[4] *Ibid.*, pp. 228-9. For other enclosures in this period see H. J. Habakkuk, *loc. cit.*, and J. D. Chambers, *Nottinghamshire in the Eighteenth Century*, pp. 147-65.

or three Acts of enclosure, but the only years that show distinct activity in Parliament are 1729-30 and 1742-3—both periods following deficient harvests and relatively high prices of food.

After the middle of the century conditions changed. Population increased, towns grew, and large-scale manufacture appeared: the prices of cereals and meat and dairy products rose, and new means of communication made it possible for land that had previously been cultivated for subsistence to grow food for the market. Circumstances were favourable to large-scale tenant farmers. After three seasons of fairly high prices of grain, the number of Acts rose for the first time, in 1754, to two figures; and after another run of poor harvests it reached thirty in the year 1759. (Thereafter, even in years of plenty, it rarely fell below this level.) In the forty-three years that followed the accession of George III, 2,000 projects of enclosure were authorised by Parliament, and of these no fewer than 1,109 were concentrated in sixteen years. Again, the periods of greatest activity generally coincided with, or followed immediately, seasons of high prices of foodstuffs (as in 1764-5, 1770-4, 1777, 1796-9, and 1802).[1]

There was a long period of relative inactivity from 1781 to 1795. The explanation sometimes offered is that the earlier enclosures had done their work and a feeling of security had replaced former anxieties as to the food supply of the nation.[2] It would be wrong to assert that the landlords were indifferent to national interests: after all, they bore the responsibilities of government. But, though arguments about the importance of staving off famine and improving the lot of the poor might count in debate at Westminster, it would be naïve to assume that the desire to enclose arose out of concern for the common good: as their attitude to the Corn Laws proves, it was not scarcity but abundance of grainstuffs that the landed classes feared. In the period 1781-95 there was only one year, and that the last of the series, when conditions of famine or near-famine appeared. On the other hand, the average price of cereals was higher than in the 'sixties and 'seventies, and if it failed to evoke a similar number of enclosures the reason must be sought in considerations of personal, rather than national, interests.

In periods of scarcity the income of farmers producing for the

[1] For statements of annual enclosure Acts see *Report on Agricultural Distress*, B.P.P. 1836, vol. VIII, Pt. 2, p. 501; *Report on Resumption of Cash Payments*, B.P.P. 1819, vol. III, p. 291.
[2] G. Talbot Griffith, *Population Problems in the Age of Malthus*, p. 176.

Agriculture and its Products

market increased, and the same was true, with a lag, of the income of landlords. But, as has already been pointed out, the expenses of enclosure were high, and after these had been met additional expenditure was usually required to put the new holdings into good shape. The large proprietor might meet this out of his own resources or, by arrangement, out of an estate in which he had a life interest. But this would usually involve a sale of assets. If the current rate of interest were low he could dispose of his holding in the Funds at a good price: if it were high he might be reluctant to sell out at the prevailing price of stock. He might seek a mortgage. When the Funds stood high he could get this at 4 per cent. or less: when they were low the rate would be 5 per cent. But he was not allowed to offer more, and when the yield on a purchase of the Funds themselves approached this level he might find it impossible to borrow at all. If most of the years of activity in enclosures followed seasons of high prices of agricultural products they were also in periods when the price of money was relatively low. In the later stages of the American war 3 per cent. stock was down in the 'fifties, and not until 1790 did it touch 80. For more than a decade rates of interest were such as to make enclosure highly expensive, and for many impossible. In 1793, when (in the early months of the year) the consolidated stock was selling at prices above 90, there was a revival of the movement; and the famine prices of grain, following the disastrous harvest of 1795, led to a spate of enclosure bills in 1796-7. It is true that government stock stood at a low figure, but these were years of high inflation, and country banks, as well as those of the metropolis, were lending freely on bond and equitable mortgage, at the legal rate of interest. The financial crisis of 1797 was followed by a recession, but renewed famine in 1800, and the fall of rates of interest that accompanied the approach of peace, brought the number of enclosure Acts to a record level in 1802.[1]

In later chapters something will be said of the fluctuations in other parts of the economy. Enclosure was thought of as part of a movement which contemporaries spoke of as 'Improvement'. At one time special attention might be paid to highways, at another to canals or buildings. High prices of grain might deflect resources from these to agriculture, and low prices might drive them back again.

---

[1] Differences of date between the beginning of harvest years, civil years, and Parliamentary sessions make it difficult to establish exact correlations. But the influence on enclosure of prices and rates of interest is clear.

Such eddies deserve study. But beneath them were tidal movements of resources of which insufficient notice has been taken. If men were quick to respond to opportunities offered by rising prices they were also sensitive to changes in costs. And, above all, their activities were determined by the degree to which funds were available for investment, in the technical sense of the word.

Enclosure related to both arable and pasture. Adam Smith believed it brought larger gains to the grazier than to the producer of cereals: 'it saves the labour of guarding the cattle, which feed better, too, when they are not liable to be disturbed by the keeper or his dog'.[1] But towards the end of the century the pressure of population on supplies of bread made it profitable to bring under the plough land that had previously been used for rough pasture: it was largely in the 'nineties, and later, that the margin of cultivation was pushed up the hillsides of the Cotswolds, the Chilterns, and other regions of thin soil. In 1796 a committee, including Sir John Sinclair, reported 'that any encouragement . . . to promote the cultivation of wastes still remaining in common, will be of little avail, unless some means are taken to facilitate their division'.[2] There is no reason why division and enclosure should have altered the acreage of either the unit of control or the unit of production; but as the hand of custom was lifted, both estates and farms could vary in size with the means and abilities of landlords and tenants, and generally the tendency was towards larger units.

The great landlords were usually well informed and able to obtain expert advice. They introduced into leases clauses designed to prevent soil exhaustion and ensure good rotations. There had been loud complaints of over-grazing of unstinted commons: this was less likely to happen on enclosed lands. The owners of estates could take long views: they thought of the soil as a continuing family asset. Even if quick returns were not to be expected they considered it worth while to sink capital in walls and drains and put up substantial farm houses, with avenues of trees, as well as solid shippens and barns. If many of them pulled down cottages (which were an eyesore, impeded husbandry, and harboured unwanted paupers), others, like the Duke of Bedford, Lord Penrhyn, John Howard, and Samuel Whitbread, built decent and pleasant houses for their labourers.[3]

[1] *The Wealth of Nations* (Everyman ed.), vol. I, p. 137.
[2] *Report of the Select Committee on the Cultivation and Improvement of Waste Lands* (1796), p. 5.
[3] Edward Smith, *The Peasants' Home*, 1760-1875 (1876), pp. 43, 61, 67, 97.

Agriculture and its Products 43

Nowhere was the long view so necessary as in arboriculture. There was growing concern at the dwindling of supplies of oak and other timber in regions sufficiently near the coast to meet the needs of the fleet. In the late seventeenth century, and throughout the wars of William and Anne, heavy calls had been made on the forests; and during the hostilities of 1739-48 there had been 'perpetual journeyings of the Purveyors of the Navy into every corner of the Kingdom to purchase timber'.[1] With each succeeding war Britain became increasingly dependent on the Baltic for supplies of the most important of all strategic materials. Acts were passed to make parishes liable for damage done to the woodlands; restrictions on the smelting of iron in coastal areas were enforced; and it was even required that the horses that brought charcoal from the woods should be muzzled to prevent their cropping the young shoots.[2] Trees, especially oaks, take long to come to full growth. The admirals who, during the final struggle with France, stuffed their pockets with acorns to plant as they took their walks in the country, were thinking of wars far ahead.[3] But the ordinary agriculturist discounted the future at too high a rate to take account of such contingencies, and, indeed, the small tenant farmer was said to have 'a natural enmity to trees and timber'. The process of enclosure, itself, might tend to deforestation. But the hedges that divided the fields helped to conserve large numbers of trees (besides giving shelter to cattle and providing screens that might be of use if ever hostile forces landed in England). It was the great landlords, on enclosed estates, who did most for the preservation of the woodlands. Whether they were thinking of the amenities of their country houses or of the revenues of their successors, it was their plantings, so it has been said, that saved Britain from naval disaster in the days before Trafalgar.

If the large estate had advantages so had the large farm. John Arbuthnot contrasted the position of a cultivator who held 300 acres, paid £150 a year in rent, and disposed of £1,500 in ready cash, with that of a man whose holding, rent, and capital were only a third of these. The first might have twelve horses and employ nine men, three boys, and three maids. In harvest time, 'two drivers, two loaders, two pitchers, two rakers, and the rest at the rick or in the barn, will dispatch double the work that the same number of hands

---

[1] E. Wade, *Proposal for Improving and Adorning the Island of Great Britain* (1755), p. 10.
[2] *Ibid.*, p. 23.   [3] R. G. Albion, *Forests and Sea Power*, p. 137.

would do, if divided into different gangs on different farms'. His overhead costs were proportionately lower: the smaller farmer would need a waggon and two carts: his larger competitor could manage with two waggons and four carts. Three hundred sheep would justify the employment of a shepherd, but not so one hundred: the smaller man allowed his flock to move at will on the common, with the result that the animals left perhaps half their wool on the bushes and furze.[1] The large farmers could afford to drain and marl the ground, and provide artificial feed for their cattle. In Norfolk 'they have converted a barren land into the most fertile spot in England. View their fields of turnips, barley, clover, and wheat, and show me such management in a country inhabited by small farmers'.

Adam Smith pointed out that the cottagers and small occupiers had some advantages: 'as the poorest family can often maintain a cat or a dog without any expense, so the poorest occupiers of land can commonly maintain a few poultry, or a sow and a few pigs, at very little. The little offals of their own table, their whey, skimmed milk, and butter-milk, supply those animals with a part of their food, and they find the rest in the neighbouring fields without doing any sensible damage to anybody.'[2] But the large dairy farm was more efficient in the production of cheese and butter, and Arbuthnot asserted that the big arable holding was best for the production of poultry and eggs: 'the threshing-floor of the great farmer is always open, and it is the corn thrown out in the straw that fattens the barn-door fowl'. Like Arthur Young in his earlier days, he could see little but good in enclosure. No doubt he overstated the social advantages of 'sensibly dividing the country among opulent men', but it can hardly be denied that the farm of 300-500 acres was generally a more economic unit than that of 100 acres or less.

The enclosures and the increase of large farms must have tended to raise the value of estates. Statistical proof is not easy, for the capital value of land depends partly on circumstances that have no connection with agriculture. Writing, in 1785, of the purchase price of land in the midlands, William Marshall said:[3]

'Some years back, the same species of frenzy—*Terra Mania*—

[1] *An Inquiry into the Prices of Provisions and the Size of Farms, with Remarks on Population*, p. 8.
[2] *The Wealth of Nations*, vol. 1, p. 207.
[3] William Marshall, *The Rural Economy of the Midland Counties*, vol. 1, p. 16. I am indebted to Dr. A. H. John for this reference.

Agriculture and its Products 45

showed itself here as it did in other districts. Forty years purchase was, then, not infrequently given. Now (1785) thirty years purchase, on a fair rental value, is esteemed a good price. There are some recent instances of land being sold, at twenty years purchase. But this may be accounted for. At the time these purchases took place, the interest of the funds was extraordinarily high. By navy and victualling bills, new loans etc., five or six per cent was made of money. And this will ever be the case. The *Interest* of the funds will always have more or less influence on the price of land. Hence those who wish to secure lands at a moderate price, should purchase when the funds are advantageous.'

The number of years purchase used as a multiplier of rents was not simply a reflection of current rates of interest: it depended also on estimates of their future course and on the prospects of improving the land.¹ But interest was the major influence. Since, in the 'nineties, rates of interest were rising, the capital value of lands tended to fall. It is not possible, therefore, to take selling prices as an index of the effects of enclosure on land values. But in another work, the writer quoted above gave his own estimate:²

'Open lands, tho' wholly appropriated, and lying well together, are of much less value, except for a sheep walk or a rabbit warren, than the same land would be in a state of suitable inclosure. If they are disjointed and intermixt in a state of common field or common meadow, their value will be reduced one third. If the common fields or meadows are what is termed Lammas land, and become common as soon as the crops are off, the depression of value may be set down at one half of what they would be worth, in well-fenced inclosures, and unincumbered with that ancient custom.'

Information on rents themselves is still more difficult to interpret.³ As already said, rents varied not only with the type and situation of pieces of land, but also with the charges, such as tithes, which

---

¹ For a discussion of the relation between interest and capital value of land at an earlier period, see H. J. Habakkuk, 'The Long-term Rate of Interest and the Price of Land in the Seventeenth Century', *Ec. Hist. Rev.* (second series), vol. v, pp. 26-45.

² William Marshall, *The Landed Property of England* (1804), p. 13.

³ The only statistical series known to the writer is one for the years 1781-1880 given in a letter to *The Times* of April 20th, 1889, and reprinted under the title, 'A Century of Land Values', *Journal of the Royal Statistical Society*, vol. LIV. (1891), pp. 528-32. It shows no clear upward movement of rents between 1781 and 1800. But the figures relate to land that came on to the market for sale by auction, and do not include whole estates. In any case the sample is too slender to bear generalisations.

the tenant might have to meet. According to Arthur Young, rents in Lincolnshire rose from 300 to 400 per cent. after enclosure;[1] but this must have been an exceptional case. In 1804 the Board of Agriculture estimated that, on the average of twenty-five counties of England, rents rose from £88 6s. per hundred acres in 1790 to £121 2s. 7d. in 1803. A large part of this increase must have been the result of a fall in the value of money; but some of it represented an enhancement of annual values in real terms.

It used to be said that, if landlords and large tenant farmers did well, the smaller cultivators suffered. It was argued that the cost of enclosure, the expenses of fencing, the arrangements in some areas for the commutation of tithes,[2] and (where rural unemployment resulted) the increase of the poor rate, bore with special weight on this class. But recent investigations suggest that the number of small farmers actually increased in the period of Parliamentary enclosure. The growth of numbers was especially marked in the case of owners with less than twenty-five acres, whose holdings, in Dr. Chambers's words, were supplementary, not basic, to their subsistence; but it was also clear in that of men with fifty acres or more, who were able to use some, at least, of the new farming techniques.[3] Under the enclosure Acts freeholders, copyholders, and tenants for lives received compensation, often in land, for the loss of their rights in the common fields. The same was true of many common-right cottagers, who had previously had no land of their own at all but were now able to buy small lots from the large proprietors, who disposed of fragments to help to defray their own expenses arising from enclosure. That the small cultivators were able to survive in face of the competition of large farmers is to be explained by the condition of the market. In the period of war and inflation after 1793 the prices of foodstuffs were such that all but the least efficient producers could show a profit: it was not until after 1815 that falling prices and agricultural depression led to a decline of small holdings.

[1] Cited by J. D. Chambers, 'Enclosure and the Small Landowner', *Ec. Hist. Rev.*, vol. X, No. 2, p. 123.

[2] V. Lavrovsky, 'Tithe Commutation as a Factor in the Gradual Decrease of Land-ownership by the English Peasantry', *Ec. Hist. Rev.*, vol. IV, pp. 273-89.

[3] E. Davies, 'The Small Landowner, 1780-1832, in the Light of the Land Tax Assessments', *Ec. Hist. Rev.*, vol. I, pp. 87-113; J. D. Chambers, 'Enclosure and the Small Landowner', *idem*, vol. X, pp. 118-27; J. D. Chambers, 'Enclosure and Labour Supply in the Industrial Revolution', *idem* (second series), vol. V, pp. 319-43; V. M. Lavrovsky, 'Parliamentary Enclosures in the County of Sussex', *idem*, vol. VII, pp. 186-208.

Agriculture and its Products

There remained the cottagers and squatters who had played little part in the open-field economy, and had no rights recognisable by the law. 'The acknowledgement of their customary though not legal privileges as worthy of compensation', wrote Gonner,[1] 'though never very frequent in the eighteenth century, grew less uncommon with the lapse of years. Early acts and awards rarely make any mention of provision for them, while at the end of the century such was far from common.' When it was made it was usually in the form of a small grant of land for the pasture of a cow, or of a fund for the purchase of fuel. There is no doubt whatever that enclosure involved injustice and hardship. (Of what large-scale agrarian reform, even in our own day, has this not been true?) The well-to-do of the eighteenth century were less sensitive than their grandchildren of the nineteenth to the sufferings of the lower orders, and it is right that historians should take account of the inequities associated with redivision of the land. Some of the cottagers who had picked up a living by casual work on the commons now had to hire themselves as labourers to farmers; some had to fall back on parish relief; and yet others left the land for the towns. There is no evidence, however, of large-scale rural unemployment. As Dr. Chambers has pointed out, the new agricultural practices (including the growing of root crops and grasses and the maintenance of large dairy herds) created new demands for labour; and the hedging and ditching now required provided employment in the winter for casual workers. There was no mass eviction: the population of agricultural villages increased at a rate not much less than that of the industrial areas; and it is not impossible that the growth of numbers was a response to increased supplies of food and greater opportunities of work on the countryside. In so far as people left for the towns the relatively high wages paid there are sufficient explanation of the movement. But the notion that poor men, like rich capitalists, might respond to opportunities of personal gain seems to arouse mental resistance: the idea that the poor were driven[2] from the land remains (and is likely to remain) firmly embedded in the text-books.

[1] E. C. K. Gonner, *op. cit.*, p. 368.

[2] So accustomed have we become to the use of this word that one able historian (who shall be nameless) has committed the absurdity of saying that 'A combination of circumstances, not the least of which was the higher and steadier wages paid in the factories, *drove* the agricultural spinner, and later the weaver, to forsake his cottage for the factory.' (Italics mine.)

For a scholarly and balanced discussion of the issues raised in these paragraphs, see J. D. Chambers's articles referred to above.

Enclosure was only one aspect of agricultural and agrarian change. The growth of large estates and large farms, the changes in land tenure, increased application of capital, technical innovations, and the growing specialisation of regions to particular crops, might perhaps have taken place without it. But if the growth of population was an independent factor—and not itself induced by these changes—it is hard to believe that, in the absence of enclosure, the standard of life could have been maintained. The dearths of 1795 and 1800 were serious enough as things were: if there had been no enclosure to increase the yield of the soil there might have been national catastrophe.

III

The chief products of agriculture were grainstuffs and livestock. Of the cereals by far the most important were wheat, which provided the staple food, and barley, which was the source of the staple drink, of most Englishmen. Rye and oats were grown to make bread or meal for the poor, as well as for all classes in the remoter parts of the country; but their chief use was as food for cattle and horses. In years of poor harvests in the second half of the century some wheat and barley were brought in from abroad; and after the middle sixties, even in normal years, Ireland supplied considerable shipments of oats.

The production of grain was stimulated and, to some extent, regulated, by government. Like rulers in all ages, the landlords who sat at Westminster, tended to identify the interest of the nation with that of the class to which they belonged. Outside opinion was rarely well informed. As Adam Smith said: 'The laws concerning corn may everywhere be compared to the laws concerning religion. The people feel themselves so much interested in what relates either to their subsistence in this life, or to their happiness in a life to come, that governments must yield to their prejudices. . . . It is upon this account, perhaps, that we so seldom find a reasonable system established with regard to either of these two capital objects.'[1] The result of self-interest and prejudice was a somewhat complicated body of legislation only the outlines of which can be sketched here.

Under the Corn Law of 1670 a sliding scale of duties had been instituted: a tax of 16s. a quarter was imposed on imports of wheat so long as the price at home did not exceed 53s. 4d.; if the price lay between this and 80s., the duty was 8s.; and if it rose above 80s.

[1] *The Wealth of Nations*, vol. II, p. 39.

Agriculture and its Products

only a nominal duty of 4d. was payable. A similar sliding scale applied to other kinds of grain brought into the country. As Mr. Lipson says, 'in view of the low level of prices during the next hundred years, the result was to give the corn producer almost complete protection.'[1]

If one object was to limit imports another was to expand exports. Under Acts passed by Parliaments of Charles II and William III, duties on the exportation of corn were removed: the merchant was allowed to engross corn for sale overseas, though similar action for sale at home was prohibited, except when prices were relatively low. Furthermore, under the Corn Bounty Act of 1688, the merchant who sent grain abroad received a present from the state of 5s. a quarter for wheat, and of smaller, but substantial, amounts for the cheaper grains. The aim was to keep the price of wheat in England at or above 48s. a quarter, or, in the words of Adam Smith, 'to raise the money price of corn as high as possible, and thereby to occasion, as much as possible, a constant dearth in the home market'. Some writers, both contemporary and modern, however, have denied that the effect was to increase prices. Mr. Lipson asserts that 'to whatever extent the bounty served to keep land under the plough, which would otherwise have gone out of cultivation, it must also have served to keep prices at a lower level'.[2] This would obviously have been true of a subsidy paid on the whole crop; but the bounty was given only on that part of it that went out of the country, and could, therefore, have no direct effect in reducing home prices. Only if it could be shown that a larger scale of output reduced marginal costs could Mr. Lipson's contention be accepted; and no one has yet demonstrated that the agricultural industry was subject to increasing returns. Mr. Lipson goes further: 'Thus', he declares, 'the bounty helped to give corn producers a greater assurance of steady and uniform prices, and consumers a better prospect of more regular supplies.'[3] But, since it operated at a fixed rate in good and bad years alike, it can hardly have influenced the fluctuation of prices. In periods of dearth, such as 1698, 1709, 1740, 1757, and 1767, the bounty was suspended and exportation prohibited; but the stabilising influence thus exerted must not be credited to the bounty itself. In fact, the bounty was partly responsible for the excessively high prices of these years. For, to quote Adam Smith again, 'The

[1] E. Lipson, *The Economic History of England*, vol. II, p. 462.
[2] *Ibid.*, p. 457.   [3] *Ibid.*, p. 458.

great exportation which it occasions in years of plenty, must frequently hinder more or less the plenty of one year from relieving the scarcity of another'.[1] The fact that in the first four decades of the century the trend of wheat prices was downward is irrelevant to the argument: prices were falling also in countries where no bounty was paid and, indeed, where the export of grain was prohibited.

Of some 43 million quarters sent abroad between 1697 and 1801, about 24 millions (nearly 55 per cent. of the whole) were exported in 1732-66. This period of thirty-four years was marked by relative stability of general prices: in 1766 the level was approximately the same as in 1715. But influences already mentioned, and others to be considered later, drove down the prices of cereals. By the middle 'sixties, however, the increased demand for white bread led to an upward movement of the price of wheat in particular; and although some 12 million quarters of cereals went abroad between 1767 and 1801, considerably larger quantities came into the country.

The change from an export to an import surplus could hardly be avoided at a time when population was growing rapidly and when England was turning from agriculture to manufacture. In several years after 1766 the bounties were suspended, exports prohibited, and foreign supplies allowed in free of duty. The facts of the situation were recognised in 1773 when a new corn law was passed under which, when prices rose above 44*s*., the bounty ceased and exports were forbidden, and when they exceeded 48*s*. the duty on imports lapsed. At the same time the bringing in of corn for re-export was freed from all duty. Adam Smith gave qualified praise to the new measure: 'With all its imperfections . . . we may perhaps say of it what was said of the laws of Solon, that, though not the best in itself, it is the best which the interests, prejudices, and temper of the times would admit of.' But his hope that it might in due time prepare the way to a better was not to be realised in his own day. In 1791 yet another measure was passed. If the Act of 1773 had assumed that the price of wheat ought to lie between 44*s*. and 48*s*., that of 1791 implied a norm between 46*s*. and 54*s*. But in each year from 1793 to 1801 the Act had to be suspended; and in the dearth of 1795-6, not only was free importation allowed, and export prohibited, but a bounty was paid on wheat brought into the country.[2]

---

[1] *Op. cit.*, vol. 2, p. 8.
[2] C. R. Fay, *The Corn Laws and Social England*, ch. III.

Agriculture and its Products

Statistics of the output of grain are few and unreliable. It is, however, beyond doubt that annual production rose considerably in the second half of the century. It is well known that the effect of enclosure was to widen the area devoted to wheat, barley, rye, and, possibly, oats. But the land turned from waste or pasture to arable does not appear to have been a high proportion of the total enclosed. It seems probable, therefore, that the growth of output of grainstuffs (and of wheat in particular) came from land already under the plough. Mr. M. K. Bennett's graphs of the yield of wheat per acre show an increase of about a third between 1750 and 1800. If his figures are even approximately correct it would appear that it was largely by more intensive cultivation, improved rotations, and better management of the soil, that England was able to supply a growing population not only with bread, but (except in years of dearth) with bread of a higher quality than had been acceptable before.[1]

Second only in importance to tillage was animal husbandry. From time to time estimates were made of the head of cattle in the country, and Arthur Young put the number of sheep at 25,589,754. (The precision of the figure implies that he either took a census or was a victim of chronic insomnia.[2]) No reliance can, in fact, be put on any of the estimates, but from 1732 a series is available of the number of cattle and sheep brought for sale to Smithfield market. It was believed, though on what grounds is unknown, that sales at Smithfield were about one quarter of those of the whole country. If the proportion of sales to the total of livestock were known it would be possible to draw conclusions as to the progress of the raising of cattle and sheep. As it is, any such attempt would be hazardous. All that can be said is that the Smithfield figures suggest substantial advance[3] and that the fluctuations they disclose are broadly what those who have made a close study of animal husbandry would expect. The prosperity of breeders and graziers must have varied with changes in the weight and quality of the animals. It used to be believed, on the authority of Sir John Sinclair and Thorold Rogers, that, over the century, the average weight of the beasts sold at

---

[1] M. K. Bennett, 'British Wheat Yield per Acre for Seven Centuries', *Economic History*, vol. III, No. 10, p. 27.

[2] The figure is cited by J. Bischoff, *History of the Woollen and Worsted Manufactures*, vol. I, p. 224.

[3] In 1732 the cattle sold at Smithfield numbered 76,210; in 1794, 109,064. The corresponding numbers for sheep were 514,700 and 717,990. For both, the increase over the sixty-two years was roughly 40 per cent.

Smithfield had doubled or trebled. But Mr. G. E. Fussell has shown that the estimates of these writers are unreliable. It seems likely that in the earlier decades the larger cattle were kept on the land, that the old-cow beef and ox-plough beef was consumed locally, and that it was only the smaller animals that were sent to London. As horses came to take the place of oxen for ploughing and carting, and as the demand for meat increased, large cattle, as well as small, would be sent to the metropolis. It is highly probable that some increase in the weight of both cattle and sheep came about (as Adam Smith insisted) as the result of increased supplies of fodder. Whether, at the same time, the quality of the meat was raised is uncertain. If the art of fattening cattle developed, the new methods of breeding associated with the name of Bakewell were confined to a relatively small number of farms; and, in any case, they were directed to a reduction of bone, and an increase of the meat, rather than to an improvement in tenderness and flavour.[1]

Small, thin cattle could be raised on the waste and the hillside, but until grain was fairly cheap the demand for meat was not large enough to warrant the setting aside of good land for pasture, and still less for the growing of roots and grass for stall-feeding in the winter. One economic determinant of the output of cattle, therefore, was the prices of the cereal crops. The raising of beef cattle takes several years: it requires more resources than that of grain. Hence another determinant is the degree of abundance of capital. In the first half of the century both prices of grain and rates of interest were falling; and in spite of the influx of Scottish cattle (or because of it, since these had to be fattened in England) English graziers seem to have extended their operations, and the same was true of the flockmasters and shepherds.

There were, however, variations in the fortunes of both. A failure of the hay crop, or a shortage of oats might make it impossible to keep livestock through the winter. Heavy slaughtering meant an immediate increase in the supplies of meat, hides, and other products, followed by a reduction which persisted until the occurrence of two or three good harvests in succession enabled the graziers to bring up their herds to full strength again. Moreover, cattle were

---

[1] Most of the statements in this paragraph have been drawn from the articles of Mr. Fussell. In particular see 'The Size of Cattle in the Eighteenth Century', *Agricultural History*, vol. III, pp. 160-81; 'Eighteenth Century Estimates of British Sheep and Wool Production' (with Constance Goodman), *ibid.*, vol. IV, pp. 131-51; and 'Animal Husbandry in Eighteenth-Century England', *ibid.*, vol. XI, pp. 97-108.

Agriculture and its Products

subject to mortality from disease. In the great famine of 1708-11 a cattle plague swept over Europe. It spread to England and reached its height here, after the famine was over, in 1714. Acting on the advice of a surgeon, Thomas Bates, the government gave orders for the slaughter of all infected beasts, and made compensation to the owners on a generous scale. The policy was successful. Within a few months the distemper had run its course: the loss of cattle (estimated at 6,000) was relatively light, and the effect on consumers and those who used cattle products for industrial purposes seems to have been negligible. For the next quarter of a century little occurred to impede the growth of the herds, though the bad harvests of 1727-9 and a drought in the summer of 1730 brought some discomfort. But the great famine of 1740 was followed immediately by a decrease in herds; and before the graziers could recover from this blow, foot-and-mouth disease once again spread from the Continent to England. It appeared in Essex in 1745, and passed rapidly to many other counties, as well as to Scotland. In spite of stringent orders for the slaughter of infected beasts, and an almost complete ban on the movement of cattle from the infected areas, it persisted for thirteen years.[1] In 1739 the number of cattle sold at Smithfield was 86,787; in 1748 it was 67,681; and not until 1760 did it reach again the level of the period before the famine. Sales of sheep also went down sharply in the early years. But after 1745, no doubt as the result of a shortage of beef, the number brought to market increased rapidly. In 1743 only 468,120 sheep were sold at Smithfield: in 1750 the figure reached 656,340. Had it not been for the relief afforded from this source the dearth of meat, hides, soap, and candles must have been acute. When, at last, in the early months of 1759 the pestilence had run its course, the King ordered that services of thanksgiving should be held at St. Paul's and at all other churches and chapels throughout the land. Thereafter, though small outbreaks occurred, as in 1781, severe quarantine regulations, and, from time to time, measures prohibiting the importation of hides and straw, saved England from a plague that seems to have been endemic on the Continent.

Sheep, also, suffered from famine and disease. Their numbers were reduced by the dearths of 1709-10, 1727-9, and 1740-1: according to William Stout, two hard winters in succession, with the

[1] C. F. Mullett, 'The Cattle Distemper in Mid-Eighteenth Century England', *Agricultural History*, vol. XX, pp. 144-65.

consequent scarcity of herbage, had by 1742 destroyed fully a half of the sheep; and in 1756-7 shortage of food again led to serious depletion of the flocks. Equally devastating was the disease known as sheep rot. Unlike the cattle plague, which was associated with drought, it was a consequence of heavy and continuous rain: its ravages were greatest among the animals pastured on the waterlogged clays of the midlands. It decimated the flocks: in 1735, it was said, the carcasses of the sheep became 'a great nuisance in the highways'. Further epidemics appeared in 1745, 1747, 1752, and, indeed, at intervals throughout the whole century. But these were usually most acute on open-field land; and, though precise information is lacking, they seem to have grown less severe with the progress of enclosure and drainage.

The raising of animals was complementary to the production of grain: most agriculturists had their own cattle and sheep. But there were large specialist graziers and flock-masters whose interests were not identical with those of the arable farmers. The relation of the price of bread to that of meat is a matter that interested Adam Smith, and might well engage the attention of present-day students of agricultural history. Little can be said about it here. Most of the statistical material on the prices of foodstuffs is drawn from the contracts made between traders and government departments or corporate bodies, such as schools and colleges. Generally these ran for a series of years: the figures do not throw much light on annual variations. It is possible, however, to use them to make one or two observations of interest. The first is that the prices of bread were less rigid than those of meat: the influence of good and bad harvests shows up more clearly than that of annual changes in the sales of livestock. A second relates to the movement of the two series over long periods of time. In the last years of the seventeenth century prices of food of all kinds were excessively high: these were known to later generations as 'the barren years'. After 1700 the price of meat fell sharply and that of bread even more so, until 1708. In 1709-10 both bread and meat rose to famine levels,[1] but from this time to 1740 there seems to have been relative stability of the ratio between the two. For the next quarter of a century, however, the price of meat was high compared with that of bread, for the losses of sheep in the early 'forties

[1] For prices in 1710 and 1711 see the list taken from the *Postbag* of December 31st, 1711, by J. E. Thorold Rogers, *A History of Agriculture and Prices in England*, vol. VII, Pt. II, p. 607.

Agriculture and its Products

and middle 'fifties were serious, and—more important—the prolonged cattle plague forced up the price of beef. The years from 1766 to 1780 showed little alteration in the ratio. But from 1781 to 1794 bread was at a lower price than in the preceding period, whereas the price of meat was rising steadily. In the dearth of 1795 the cost of meat rose even more steeply than that of bread, and the same was true to the end of the century.[1]

It has already been said that in the years after 1780, enclosure was directed specially to land that could be turned to the plough. No doubt the increased demand for meat—a sign of a rise in the standard of living—was partly responsible for the increased prices of beef and mutton. But the deflection of land to arable must also have played a part. In the last years of the century when wheat prices were going up at an alarming pace, there were not wanting voices to protest that too much land was still under grass. It was not merely cattle and sheep that were the culprits: large resources were being absorbed in the raising of horses. These had largely taken the place of oxen at the plough; the numbers on the roads had increased; and, if fewer were now employed to turn machines in factories, the demands of the hunting field and the army were insistent. The horse required a good deal of land for pasture: his daily food, according to a writer of 1799, consisted of three pecks of oats, a gallon of beans, and an unspecified quantity of hay. He was 'the most dangerous moth in the whole web of agricultural economy'.[2] The demand for land as pasture, and for the growing of oats, beans, and other fodder for horses, may not have been as important as it appeared to this writer; but it must not be ignored in any discussion of the causes of the high price of grain at this time.

IV

Agriculture provided materials for a number of industries. Wheat passed from the farms to the corn millers, bakers, distillers, and starch makers: in the form of flour it was used by stationers bookbinders, linen printers, trunk-makers, and paper-hangers, Barley also went to the distilleries, but its chief industrial use was to make malt and, hence, beer and ale; and rye meal was of some importance in tanning.[3] Sheep were the source of the material of the largest

---

[1] See Lord Beveridge and Others, *Prices and Wages in England*, vol. I.
[2] W. Tatham, *The Political Economy of Inland Navigation* (1799), pp. 412-13.
[3] Charles Smith, *Tracts on the Corn Trade and the Corn Laws* (ed. 1804), p. 224.

of English textile industries, and cattle provided hides for tanners and hence leather for a variety of trades. The fat of both was used by soap-boilers and candle-makers, the horns by cutlers, and the bones by the glue manufacturers. The list could be extended. Since most of the industries that worked up the products of agriculture were subject to excise duties, there is statistical information as to output and to the variations of activity over most of the century.

Milling and baking were carried on in all parts of the country, but there was a growing tendency towards concentration in centres of communication, and especially in London. In earlier times the miller had ground corn brought to him by farmers, dealers, and householders, and had received for his labour a fee which, by ancient custom, was equivalent to every twentieth grain (or 5 per cent. of the corn).[1] During the earlier part of the eighteenth century, however, as the laws against engrossing and regrating fell into disuse, millers and bakers tended to buy their own raw materials. They were thus able to keep their machines and labour more regularly employed. But in assuming some of the functions of broggers and mealmen they became subject to the odium attached to all who made profit by dealing in the necessaries of life.

Generally the miller could make whatever kind of flour he chose and charge what he could get. But the position of the baker was less enviable, for in many of the larger towns the assize of bread was still operative. In earlier times the price of the loaf had been fixed according to the price of wheat. But after 1710 the magistrates were allowed to take into account the price of flour, and in 1758 the number of loaves to be made from a sack was prescribed by law.[2] At the same time flour was divided into two grades, the best for making wheaten bread, the second for what was called the household loaf. The authorities had the difficult task of fixing the price of a commodity the content of which might vary from one baker to another. Until 1797 they had discretionary powers as to the allowance to cover the cost of baking; but in this year the remuneration of the baker was fixed at so much a sack of flour turned into bread.

[1] *Report of the Committee on the Methods practised in making Flour from Wheat* (1774) p. 12.
[2] C. R. Fay, *The Corn Laws and Social England* p. 46. In 1792 the London magistrates fixed the price of bread by reference to the price of flour, and ignored that of wheat. Luke Heslop, *Observations . . . concerning the Assize of Bread* (1799), p. 5.

Agriculture and its Products

That there was ample opportunity for malpractice is obvious. Millers and bakers were accused of using inferior materials, of putting the flour through bolting cloths of unduly wide mesh, and of mixing the better with the inferior grade in making the wheaten loaf. It was perhaps because of such possibilities that the prices of bread were normally below those fixed by the magistrates. There is, however, no reason to think that millers and bakers had more than their share of original sin. That many of them made high profits in time of scarcity is true: a man who had bought stocks of grain or flour beforehand would benefit by the rise of the assize price of bread. But an examination of the course of prices of wheat, flour, and bread at various points of time gives no ground for believing that there was widespread exploitation of the public.[1] In periods of scarcity the price of flour rose less than that of wheat, and the price of bread less than the price of flour. (This is what one would expect, for the other costs of the millers and bakers were not increased at such times.) Many of the millers were Quakers and it may perhaps be assumed that their conduct did not depart seriously from the standards laid down by their Society. Nevertheless, in times of dearth they and the bakers were liable to have their mills and ovens destroyed by angry mobs, and their lives were often in danger.

The trends of prices of bread, then, followed closely those of wheat. Both were falling in the 'thirties, and almost stationary from then to the middle 'sixties. Thereafter prices rose, at first moderately, but, in the last decade, steeply. The lowest price was in 1743 and 1744, the highest in 1800.

A large part of the wheat crop each year went to the distilleries. Something has already been said of the vast expansion of the output of British spirits in the first half of the century: in 1751 duty was paid on over 7 million gallons. After the Act of that year, however, the trend of production was strongly downward, and from the early 'sixties to the middle 'eighties output was roughly stable at between 2 and 3 million gallons. But a reduction of the duty in 1785 inaugurated a new period of relatively high activity, and during the 'nineties outputs of between 4 and 5 million were common. The industry was subject to violent fluctuations: the years in which production rose to outstanding heights were associated with abundant harvests, and those that registered abnormally low levels coincided with, or

[1] The basis of this statement is the table in Appendix D of the Report of 1774, pp. 24f.

followed, seasons of dearth.[1] In some years of acute scarcity of wheat, the use of corn for distilling was prohibited, and output fell almost to zero.

Yet another part of the wheat harvest went to the making of starch, a commodity needed not only by housewives and laundresses but by hairdressers, paper-makers, and textile workers (who used it in preparing their yarn for the loom). The manufacture of starch, like distilling, was largely concentrated in London, and the two competed for wheat in the same market. During the first half of the century the output of starch fell as that of spirits rose, and by the middle of the 'thirties had sunk to a level only about a half that of 1720: the age of gin was an age of dirt and untidiness. The Acts of 1743 and 1751, however, led to a reversal of these movements: the production of starch increased just as rapidly as that of spirits declined. In the 'seventies, stimulated, no doubt, by the growing demand of the textile industries, the supply of starch increased sharply, while that of spirits remained stable; but the imposition of heavier duties on starch in 1781 and 1783 checked the movement, and released wheat for the distilleries once more. In the dearth of 1795 manufacturers of starch, hair-powder, and blue were prohibited from making use of wheat or other foodstuffs, and a similar measure was promulgated at the height of the famine in 1800.[2]

Barley could be turned into spirits and starch but its chief industrial use was to produce malt. The output of the malt-houses, as measured by the excise figures, shows no such long-term oscillations as that of the distilleries. After the famine years of 1709-10 the trend was upward to 1721, and after that for a whole century, slightly downward. In times of good harvests more barley went to make drink, and, in seasons when yields were low, malting barleys might be used for bread. Sometimes the crop was fairly good when the wheat crop was deficient, but usually the output of both grains varied in the same direction. This was not so of another ingredient in beer: the hop vines were liable to a disease known as blast, which affected the crop in alternate years and produced curious annual fluctuations of output between 1766 and 1781, in particular. As an element in the cost of brewing, however, hops were of far less account than

---

For changes in the duties see *Report of the Commissioners of Inland Revenue* (1870), vol. I, pp. 57-8, and for figures of output *ibid.*, vol. II, pp. 4-6.

[2] For the duties on starch see *Report of the Commissioners of Excise* (1834), p. xxv.

Agriculture and its Products

barley—even though twice the quantity used at the present day normally went to the standard barrel.[1]

Beer is a beverage of many varieties and strengths. In the first three-quarters of the century the increase of production of strong beer (or ale) was slight, but, in the last quarter it was substantial. (In 1701 the output was 3·4, in 1775, 3·9, and in 1799, 5·8 million gallons.) The statement sometimes made that people were forced to take to the pernicious practice of drinking tea by a growing shortage of malt liquors is palpably false. Unlike malt, beer is a finished product. The materials of which it was made could be stored from one season to the next; and, since time was needed to malt the barley and dry the hops before brewing could begin, the effects of the harvest on the output of beer might not be registered for several months. Partly for this reason, the course of production is marked less by sudden changes than by wave-like movements each extending over a number of years. Most harvest labourers received part of their payment in beer, and most industrial workers spent a significant part of their wages on it. The fluctuations of output have a bearing, therefore, on the standard of comfort of labour. Good crops stimulated the activity of bakers and brewers and brought general well-being: bad crops depressed production of bread and ale and increased poverty.

The prosperity of other manufactures varied with the supply of animal products. The greatest industry of all was based on wool that came largely from British sheep. But imports from Spain and other countries were of some importance, and though depressions in years of dearth, like 1728 and 1740, may be attributed to deficiencies in the supply of wool, other forces arising from the state of the market for cloth, at home and abroad, were probably of greater moment. Tallow from sheep was the chief raw material of the candle makers, and though many people still made their own tallow dips (and so evaded the tax) the excise returns are an index of the activity of a not-unimportant body of specialised workers. The industry was carried on chiefly in London, and the make of tallow candles for the period 1732-70 shows a remarkably close correlation with the number of sheep sold at Smithfield. (Years of hunger, such as 1728-9 and 1740-1, must have been years of darkness also for many poor people.) Between 1715 and the end of the century the output of

[1] For the oscillations of the output of hops see J. Bannister, *Synopsis of Husbandry* (1799), p. 202, and H. H. Parker, *The Hop Industry*, p. 42.

tallow candles doubled, and that of wax candles, which were used
by the rich, increased nearly tenfold. Much the same thing occurred
in the case of soap, which was made with fat obtained from the
butchers. The upward trend was broken in the late 'twenties, and
there was contraction and stagnation during most of the years of the
cattle disease. After 1775 output increased rapidly with rising demand
from the textile industries, and again over the whole period 1715
to 1800 there was a doubling of output. Finally, the tanning of leather
(and perhaps, therefore, the making of footwear and saddlery) shows
variations, most of which coincided closely with those mentioned
above. There was the same fall in the late 'twenties and the early
'forties, and the influence of the cattle plague is again visible. This
is not surprising since the industry was localised in the metropolis
and depended on the trade of Smithfield. Most of the tanneries were
on the banks of streams entering the Thames on the Surrey side.
They needed large quantities of water for cleaning the hides as well
as for use in the tan pits; and hence output was affected by the
recurrence of droughts and frosts. Nevertheless, the annual variations, as well as the trends, seem to have been determined mainly
by the ups and downs of the supplies of hides and skins.

Enough has been said to indicate the bearing of agriculture on
the manufactures that made use of its products. But it was not only
this group of industries that was affected by the condition of agriculture. The ploughing of the soil and sowing of seed were an act
of investment, in the economists' sense of the word. Months later,
when the harvest appeared, further investment was called for and
the magnitude of this depended on the size of the crop. If the harvest were abundant an army of men, women, and children had to
be recruited to mow and glean, cart, store, and winnow the grain.
Incomes were paid out which were likely to be spent more largely
on other things than food, since, in such conditions, this was cheap.
A stimulus was given to industries concerned with consumers' goods,
and this was transmitted throughout the economic system. Increased
demand led to an extension of employment and a rise of money
wages. Of the harvest year 1730-1, William Stout of Lancaster writes:[1]
'All provisions, plenty and cheap; and the linen, woollen and iron
manufactures sell well, but flax, wool and iron cheap, spinning (wages)
advanced one-fourth, so that labouring people may live comfortably,
which they should remember.' And of the following year, when the

[1] *Autobiography of William Stout of Lancaster* (1851).

crop was even better, he reported that, 'corn proved good and is very low, and also provisions of all sorts cheap; and our linen, woollen and iron manufactories sell well, and spinning is at the highest'. A generation later Adam Smith expressed the connection between good harvests and good wages in other terms.[1] 'In years of plenty', he observed, 'servants frequently leave their masters, and trust their subsistence to what they can make by their own industry. But the same cheapness of provisions, by increasing the fund which is destined for the maintenance of servants, encourages masters, farmers especially, to employ a greater number. . . . The demand for servants increases, while the number who offer to supply that demand diminishes. The price of labour, therefore, frequently rises in cheap years.' He adds that 'cheap years tend to increase the proportion of independent workmen to journeymen and servants of all kinds . . .' In face of this and much other evidence to the same effect, it is strange, indeed, that the most commonly held theory of wages was that they depended on the cost of subsistence.

There were other effects of abundant harvests. During the first sixty years of the eighteenth century Britain was, generally, a net exporter of grain. In years of abundance the amount of cereals exported might be such as to lead to a favourable balance of trade and sometimes to a net import of bullion. Whether through the effect of this on credit, or, more directly, through its effect on the incomes of exporters and others, there may well have been a tendency, at such times, to a rise of earnings and employment throughout the community.

A bad harvest, especially when it was followed by a second, might have dire results. The investment called for was small. Most people had to cut down their purchases of manufactured products, and so unemployment spread and wages fell. 'More people want employment than can easily get it', wrote Adam Smith,[2] 'many are willing to take it upon lower terms than ordinary, and the wages of both servants and journeymen frequently sink in dear years.' In 1727-8 William Stout reports 'a very sickly summer, and a great mortality in the plain country. Corn proved dear. . . . The linen manufactory very low, and spinning one third less than last year; so that the poor have a hard year.' Similar conditions prevailed in the dearth of 1739-40: according to Adam Smith, 'many people were willing to work for bare subsistence', and Stout records, 'great complaints of

[1] *The Wealth of Nations*, vol. I, pp. 74-5.   [2] *Op. cit.*, p. 75.

the sale of our woollen, linen and iron manufactories; and spinning fallen from 9*d*. to 5*d*.'. 'Cotton wool', he adds, 'will not sell, and linen yarn very slowly, so that poor labouring people are much straitened to live; also the wages of servants at Whitsuntide came down, many endeavouring to live without servants who used to have (them).' In times like these it was usual to prohibit the export of grain and suspend the duties on import. These measures may have given immediate relief, but they also tended to turn the exchanges against Britain, to lead to a restriction of credit, and to accelerate the fall of purchasing power. Among the causes of instability of economic life in this century variations in the yield of the soil must be given first place.

CHAPTER THREE

Internal Trade and Transport

I

'THE home trade', wrote David Macpherson, 'is with good reason believed to be a vast deal greater in value than the whole of the foreign trade, *the people of Great Britain being the best customers to the manufacturers and traders of Great Britain.*' In a footnote he adds that 'it has been calculated, though I know not upon what grounds, or whether it is at all calculable, that the home consumption of this country is *two-and-thirty times* as much as the exports to foreign countries'.[1] The character of this internal trade varied with the nature of the commodities concerned and the degree of public control exercised: it is impossible in a work on this scale to offer more than a general and imperfect sketch of its features.

Early forms of organisation persisted. Stourbridge Fair at Cambridge was still a national event, and the traders who assembled there once a year dealt in goods of all kinds. Most of the other fairs, however, had come to specialise in the staple products of the region in which they were held. Such, for example, was so of those held at Norwich for cattle, Northampton for horses, Weyhill and Burford (in Dorset) for sheep, Worcester for hops, Yarmouth for fish, Colchester for oysters, and Ipswich for butter and cheese. But, as population thickened and means of communication improved, weekly markets, at several hundred towns in England and Wales, took over much of the trade that had previously been done at the fairs. These markets tended to specialise increasingly in agricultural or semi-manufactured products. There were markets for grain at Maidstone, Ipswich, Lynn, York, Ormskirk, Gloucester, and Devizes, and for wool at Cirencester, Leicester, and Lincoln. Yarn and cloth were disposed of at markets in many towns of East Anglia, Yorkshire, and the south-west—often on stalls in the streets, but later in the solid, and sometimes gracious, cloth halls that were built at places like

[1] D. Macpherson, *Annals of Commerce*, vol. iii, p. 340.

Colchester, Ipswich, Witney, Wakefield, Bradford, Halifax, and Huddersfield.

Fairs and markets served the needs of their own areas, but were also points of assembly for the traders of the metropolis. This was true, in particular, of the ring of markets at places not more than fifty or sixty miles from the capital: Farnham, Guildford, Henley, and Reading specialised in wheat, Abingdon, Hertford, and Reading in barley, and Dorking in poultry. Grain brought down the Thames and the Lea was sold at Queenhithe, and other supplies carried round the coast were dealt in at Bear Key; but, in the period of civic improvement in the 'sixties, the market for grain was concentrated at the new Corn Exchange in Mark Lane. Fodder from Middlesex and other neighbouring areas was sold on the pavement at Haymarket and Whitechapel, livestock from the grazing counties at Smithfield, and cheese from Cheshire, Suffolk, and Gloucestershire, along with meat and poultry, at Leadenhall. The main market for fish was at Billingsgate, that for vegetables and fruit at Covent Garden. Over most of the century, dealings in coal took place on open ground near the wharves, but in 1768 a Coal Exchange was erected in Thames Street. And throughout, though of diminishing importance, Blackwell Hall remained the central market for cloth.[1]

London was not only a collecting but also a distributing centre. A large part of the wheat, barley, and livestock brought there was returned to the provinces in the form of flour, meal, starch, spirits, beer, leather, tallow, and so on. But it would be wrong to give the impression that all roads led to and from London. Traders moved about the country, buying produce here and disposing of it there as opportunities arose. They had various designations and duties. Dealers or factors took grain from the farmers and sold it to millers; these handed it on, in altered shape, to mealmen who supplied the bakers with the mixture they required. Graziers, whose business was to fatten cattle, bought from breeders, employed drovers, and transferred the beasts to market salesmen; and these supplied carcasses to wholesale butchers who, in turn, met the needs of the retail cutting-butchers. Coal from the north was dealt with successively by Tyneside fitters, shipmasters, London crimps, wholesale dealers, and retailers or hawkers. Wool was passed on from the flockmasters

[1] For these and many other relevant details see R. B. Westerfield, 'Middlemen in English Business, particularly between 1660 and 1760', *Transactions of the Connecticut Academy of Arts and Sciences*, vol. 19, *passim*. See also D. Defoe, the various editions of *A Tour through England and Wales*.

to wool-staplers who sorted the fibres and provided the clothiers with a wide range of choice. After the wool had been spun it might be dealt in by yarn merchants and, after weaving and other processes, passed to drapers and shopkeepers.[1] But the links in each of these chains varied from time to time: distribution never followed a set pattern.

Public sentiment and legislation had some effect on the organisation of trade. Generally they were hostile to the man who disposed of things without altering their form. 'We have a wise law which says no man shall sell beef who does not kill beef', wrote Charles Townshend,[2] and there was a similar injunction that forbade the miller to sell corn, except after it had been turned into flour, meal, or malt. But there was also a strong feeling against those who meddled with other men's business. Townshend looked with a friendly eye on a law of a different tenor from that mentioned above: it was that 'no one who fats cattle shall follow the business of a butcher'. According to the same rule, it was held that drovers, whose business was transport, ought not to act as dealers, and that the 'common brewer', the importer of wine, and the distiller of spirits should not sell by retail.[3]

Such laws and customs were, however, increasingly disregarded. The new capital and enterprise that poured into the inland trade flowed largely outside the channels prescribed by law or tradition. The distinction between dealer and factor was blurred: at one time a man might buy grain or cattle outright; at another he might act as an agent (whether for the farmer or the wholesaler) and receive a commission. A grazier might either buy in the market or intercept the drover and strike his bargain on the road. A miller might turn jobber or mealman and draw as much of his income from a difference in prices as from his own authorised calling. London fishmongers took shares in fishing smacks; early in the century the lightermen of the river became crimps or coal factors; and the hoymen of Kent, whose original occupation was shipping, obtained stalls in the Corn Exchange in Mark Lane. On the other hand, the grain

[1] R. B. Westerfield, *op. cit., passim.*
[2] *National Thoughts recommended to the serious attention of the public* (n.d.), p. 12.
[3] The 'brewer victuallers', however, who flourished in the country towns might also act as publicans. In London and some other cities the competition of the large-scale brewers of porter forced most of them to cease production and become retailers. Peter Mathias, 'Revolution in Brewing', *Explorations in Entrepreneurial History*, vol. V (1953), pp. 211-12.

dealers of London became carriers: about 1760, it was said that they owned all the barges at Ware. With or without licence, tallow chandlers sold spirits, and, after 1750, brewers began to buy up ale-houses and control the retail trade over wide areas.'[1]

There was a tendency, observable in the early years of the century, to deal in grain, cheese, and other produce on the basis of samples. This made it easier to buy in one market and sell in another within a short space of time, and to enter into contracts for future sale and delivery. Since 1663 engrossing of cereals had ceased to be illegal so long as prices were kept below specified levels. The result was the appearance, in the open, of jobbers or speculators who never set eyes on the things they dealt in. Sometimes, whether by law or by rules of their own determining, dealers were able to limit their numbers, restrict supplies, and force up prices. In 1772 it was observed that the London cheesemongers had a club at which they settled each week the amount of produce each member might bring to the city.[2] The coal trade was notorious for its combinations: throughout most of the century the fitters of the Tyne regulated the vend and determined quotas. At the London end of the trade the lightermen buyers exercised a similar control: as early as 1729 ten of these, it was said, handled two-thirds of the trade and held regular meetings to settle prices. In 1788 there were only twenty-seven 'first buyers' on the London Coal Exchange, and these had an agreement as to the number of chaldrons that might be unloaded from the collier vessels each day.[3] Less is known (though much has been written) about the operations of the grain dealers, but they, too, were inclined to restriction: towards the end of the century the number of factors on the London Corn Exchange had been reduced to fourteen.[4]

Notwithstanding this evidence of the existence of privileged positions, it seems probable that the changes in the structure of distribution extended competition. It was not, it is true, until the nineteenth century that the sovereignty of the consumer was clearly proclaimed; but the increased investment in domestic trade at this earlier period did much to reduce local monopolies. The growing use of the services of factors, by both buyers and sellers, speeded the exchange of goods—for the factor (paid as he was by commission)

---

[1] R. B. Westerfield, *op. cit.*, pp. 154, 166, 179.   [2] *Ibid.*, p. 208.
[3] T. S. Ashton and J. Sykes, *The Coal Industry of the Eighteenth Century*, p. 204.
[4] R. B. Westerfield, *op. cit.*, p. 154.

Internal Trade and Transport

had an interest in seeing that the volume of transactions was high and that no time was wasted in unnecessary bargaining. It is not easy to disentangle the results of improved marketing from those of improved transport. In 1722 Defoe pointed out that in earlier times farmers had been obliged to sell off their cattle in the autumn because the roads were impassable in bad weather: when the highways near London were improved it became possible to market livestock at all times of the year, and hence to assimilate winter and summer prices.[1] In the same manner, after 1760, when the colliers from the Tyne began to make voyages the year round, the differences in the prices of coal between one season and another lessened. But in both cases something is to be attributed to a simultaneous development of middlemen, who provided storage facilities, and so helped to iron out price differences over periods of time. The professional trader was ready to buy or sell on request. Corn dealers bought largely in the autumn, and so provided farmers with money to pay their Michaelmas rents and hire labour: their sales were greatest in the late months of the harvest year, when prices would otherwise have been driven up. Grain could be stored for several years. Farmers could hardly have carried large stocks, but the corn dealers had resources that enabled them to hire granaries not only in England, but also overseas. The cost of storage obviously varied with the rate of interest, and since this was generally lower in Amsterdam than in London, it was not unusual for the English corn merchants to hold part of their stocks in Holland. By doing so they obtained the further advantage of drawing the bounty on the export of grain: in times of dearth at home they could usually bring back their holdings without payment of duty.[2]

The durable and highly diversified products of manufacture were less suited to handling by chains of middlemen or to sale in public markets. In the early years of the Coalbrookdale enterprise Abraham Darby disposed of his pots, kettles, and other wares through the annual fairs at Chester, Wrexham, and Stourbridge (Worcestershire). But though his successors continued to attend these fairs to obtain orders or receive payment for deliveries made during the previous twelve months, the bulk of their output was disposed of through the merchant house of their partner, Thomas Goldney, at Bristol.[3] From this, no doubt, some of the bar iron, as well as the smaller

[1] Cited by R. B. Westerfield, *op. cit.*, p. 193.   [2] *Ibid.*, p. 163.
[3] A. Raistrick, *Dynasty of Iron Founders*, p. 7.

castings, passed to other iron merchants, whose business it was to have the bars slit and to sell rods of various lengths and breadths to manufacturers and craftsmen. The makers of tools, locks, hinges, cutlery, and nails often disposed of their products to wholesale ironmongers, who had warehouses and employed riders-out to seek orders from shopkeepers and others. A similar organisation existed in the textile trades where the wholesale draper performed services like those of the ironmonger. Increasingly, however, manufacturers of non-perishable goods, like Peter Stubs, file-maker of Warrington, made use of trade-marks and other devices, to establish direct dealings with retailers. The elimination of middlemen in the sale of manufactured goods was, indeed, one of the outstanding features of the rise of large-scale industry.[1]

The history of shopkeeping has yet to be written. In the early part of the century some retailers, especially those dealing in food, were content with a stall in the market, a shed in the street or against the wall of a church, or even a cellar basement. Craftsmen often used the ground floor of their houses as shops and worked at their trade in a room at the back or upstairs. In the rebuilding after the Great Fire some houses were specially designed for retailing, and the new Royal Exchange was fitted out with rows of shops, each with its own sign.[2] The author of an early essay on the drapery trade had urged that lest 'commodities be sold too deare, shops shall not be too darke; and lest they be sold too cheape, they shall not be too light'.[3] But the eighteenth century stood for enlightenment, and the balance was tipped towards a clear exhibition of wares. In its early years only the better shops had glass windows: the others had simply apertures that could be boarded up at night. When William Stout served his apprenticeship to a retailer at Lancaster it was in a shop of this kind that he worked, exposed to the weather, summer and winter alike.[4] Sometimes the window-board was hinged so that, when it was down, it served as a shelf projecting into the street. As time went on, this became a fixture and was used as the base for a glass window which extended to the cornice of the overhanging first floor. In London regulations forbade such projections

[1] T. S. Ashton, *An Eighteenth Century Industrialist*, ch. V.
[2] R. B. Westerfield, *op. cit.*, p. 348.
[3] William Scott, *Essay on drapery or the complete citizen* (1635). Cited by A. H. Cole in his Introduction to the reprint of *Thomas Watts, An Essay on the Proper Method for Forming the Man of Business* (1716).
[4] William Stout, *loc. cit.*

Internal Trade and Transport

for more than ten or eleven inches beyond the line of the foundations. But within these limits they were legalised for shops in 1772. The gently curved bow-window was no serious impediment to the passer-by, and it brought a new grace to the narrow streets of the capital.[1]

Much of the retail trade of the large towns was done by hawkers: the streets of London rang with the cries of women offering hot bread, gingerbread, apples, watercress, fish, rabbits, and cats' meat.[2] The licensed chapmen and pedlars still served the needs of the rural area, but as population grew many of these began to take shops, some of which opened only on market days but which served as stores for the wares they hawked in the surrounding villages. Until the later decades of the century shopkeepers in the provinces seem to have dealt in a wide range of goods: there was little of that specialisation to a particular commodity that existed in the wholesale trade. Stout's master, Henry Coward, was described as a grocer and ironmonger: his stock included sugar, dried fruit, tobacco, brandy, hardware, and nails. When Stout himself set up in business he dealt in a similar miscellany: cheese, soap, ginger, cutlery, and copperas (used in dyeing), as well as the commodities just referred to, are mentioned in his records. Peter Davenport Finney, who began as a confectioner in Manchester about 1755, soon added a large grocery establishment to his business. Both Stout and Finney became wholesalers, without, however, relinquishing their retail trade, and both took shares in vessels and ventured into overseas commerce. There was no foolish idea that retailing was a less worthy calling than merchanting: Finney came of landed gentry, and after making a fortune, mainly out of sugar and treacle, retired, in 1768, to the family estate at Fulshaw.[3] Other instances are to hand of the association of retail with wholesale trade: Peter Stubs was, at the same time, an innkeeper, a file-manufacturer, and a dealer in tools and other wares; and some of the other wholesalers from whom he drew supplies had their own retail shops.

It was not difficult to set up in business as a shopkeeper. Wholesalers were usually willing to supply credit. When Stout went to London to get his first stock in 1687 he obtained goods to the value of £200 but was required to pay only half in ready money, 'as was

---

[1] William Hoskins, *Guide to the Regulation of Buildings in Towns* (1848), pp. 51-3.
[2] M. D. George, *op. cit.*, p. 159, and R. B. Westerfield, *op. cit.*, p. 317.
[3] Samuel Finney, *An Historical Survey of the Parish of Wilmslow*. MS. in the possession of Dr. Finney of Wilmslow.

then usual to do by any young man beginning trade'; and a century later a writer who disliked small retailers attributed the increase in their numbers to the abundance of 'low credit'.[1] The trade directories of the later decades bear witness to the multiplication of shops. They give evidence also of a growing specialisation of retail trade and of a change in policy respecting prices. At the beginning of the century, as in large parts of the East today, dealings between retailers and customers had been accompanied by much haggling. William Stout, like most of his fellow Quakers, was opposed to the practice. He usually 'set the price at one word, which seemed offensive to many who think they never buy cheap except they get abatement of the first price set upon them'. Haggling, however, arose not only from cupidity but from the fact that, in the absence of daily or weekly information of the state of the wholesale market, neither shopkeeper nor customer knew what was the appropriate price at any point of time. As markets widened, as price-lists appeared, and as advertisements in newspapers spread knowledge of what wholesalers were charging, retailers ceased to concern themselves greatly with determining prices: they became content to receive a more or less fixed percentage return on their outlays. By the end of the century the habit of setting a price-ticket on goods was extending; and, increasingly, retailers competed in the range of choice offered to their customers, in terms of credit, and in advice and service, rather than in prices.

II

Throughout the eighteenth century the chief highway of the English was the sea. Beside the ships of the Royal Navy and the merchantmen engaged in overseas commerce, large numbers of small craft trafficked in the waters about Britain. 'There are supposed to be about eighteen hundred ships and vessels in the Coal trade and about nine hundred more in what they call the Northern trade', wrote a naval officer[2] in 1774. These plied along the eastern shore between the Scottish ports, Newcastle, Hull, Lynn, Yarmouth, and London. In the west 'from the ports of Whitehaven, Milford, Swansea, Neath, Barry and Carmarthen . . . there cannot be less than

[1] Charles Townshend, *op. cit.*, p. 9.
[2] Robert Tomlinson, 'Essay on Manning the Royal Navy,' *Pub. Navy Records Society*, vol. LXXIV, p. 129 and p. 130. For other estimates of the number of colliers, see Ashton and Sykes, *op. cit.*, p. 200.

half as many vessels employed as from the northern ports', though these, it was added, were of lesser tonnage. The principal cargoes consisted of coal, stone, slate, clay, and grain—all commodities the weight of which in relation to their value made the cost of carriage by road prohibitive. Horsehay in Shropshire cannot be more than fifty or sixty miles by road from Chester, and the country between the two is almost unbroken plain. Yet, as late as 1775, it paid the Horsehay Company to deliver its pig iron to Chester by carting it to the Severn and thence transporting it by trows to Bristol, where an agent, John James, put it on a vessel that sailed round the Welsh coast to the Dee.[1]

It was not, however, only bulky and heavy goods that were carried circuitously by sea. At the beginning of the century the retailer, William Stout, brought his groceries from London to Lancaster by coasting vessels,[2] and a hundred years later, according to a Yorkshire merchant, the less valuable kinds of wool went from London to Leeds mainly by sea.[3] Adam Smith,[4] indeed, asserted roundly that goods of all kinds could be carried more cheaply by salt-water than by road:

'Six or eight men . . . by the help of water-carriage, can carry and bring back in the same time the same quantity of goods between London and Edinburgh, as fifty broad-wheeled waggons, attended by a hundred men, and drawn by four hundred horses. Upon two hundred tons of goods, therefore, carried by the cheapest land carriage from London to Edinburgh, there must be charged the maintenance of a hundred men for three weeks, and both the maintenance, and, what is nearly equal to the maintenance, the wear and tear of four hundred horses as well as of fifty great waggons. Whereas, upon the same quantity of goods carried by water, there is to be charged only the maintenance of six or eight men, and the wear and tear of a ship of two hundred tons burden, together with the value of the superior risk, or the difference of the insurance between land and water carriage.'

In a later passage[5] he remarks that 'Live cattle are perhaps the only commodity of which the transportation is more expensive by

---

[1] Records of the Horsehay Co. Waste Book C., 1774-81.
[2] *Autobiography of William Stout.*
[3] *Minutes of Evidence relating to Wool:* evidence of William Hustler, April 29th, 1800. B.P.P., vol. XL, p. 6.
[4] *The Wealth of Nations,* vol. I, pp. 16-17.   [5] *Ibid.,* vol. I, p. 403.

sea than by land'—for the very good reason that 'by land they carry themselves to market'.

Carriage by sea was, however, liable to interruption and delay. Until 1760 the collier fleets stayed at their moorings during December and January;[1] and even at other seasons adverse winds might hold up shipping in the Tyne or Thames for weeks on end. One reason why the finer woollens went by land was the 'detention and injuries' to which they were subject when carried by sea. In time of war men from the coasting vessels were swept into the navy, and French or Spanish privateers infested the sea lanes around Britain. But it was not only tempestuous seas and the King's enemies that preyed on the traffic: the men who loaded and unloaded the cargoes took at least as heavy a toll. Patrick Colquhoun (not, it is true, the most exact of statisticians) put the annual losses from pilfering on the Thames at no less than £506,000.[2] The rulers of England were supposed to look with special favour on a calling that bred men for His Majesty's ships. But even they put barriers in the way of the coastwise trade: coal carried to London by sea was subject to a duty equal, for most grades, to 60 per cent. of the price at the collieries.[3] Such varied impediments must have led many traders to seek alternative means of reaching their markets. If one reason for their interest in the improvement of inland communications was to obtain better access to the ports, another was a wish to avoid the hazards and delay involved in transport by sea.

III

In their natural state few of the English rivers were well suited to navigation, and from early times the erection of dams by cornmillers, and of garths by fishermen, had raised artificial obstacles to their use for transport. The barge, moving up or down stream, had often to wait for hours, or even days, till the miller could be persuaded, or bribed, to raise the middle timbers of his weir, and so give the 'flash' that brought the waters, above and below, to roughly the same level. 'The Severn', Dr. Willan[4] points out, 'was remark-

[1] *Report on the Coal Trade* (1800), Appendix, p. 553.
[2] P. Colquhoun, *The River Police* (1800), p. 154. The figure includes losses from vessels in overseas as well as domestic trade.
[3] A. Smith, *op. cit.*, vol. II, p. 356.
[4] T. S. Willan, 'The River Navigation and Trade of the Severn Valley, 1600-1750', *Ec. Hist. Rev. VIII*, No. i, pp. 68-9.

able as being the only great English river that could be navigated without flashes, floodgates, locks or sluices.' Though it was subject to floods in its higher reaches, it was navigable at most times from Bristol to Welshpool; and its tributaries brought coal from Shropshire, salt from Droitwich, and grain from several counties, as return-cargoes for the trows that carried tropical and other produce upstream. The Thames, on the other hand, presented serious barriers to the navigator: though it served to carry corn and other products to market centres such as Oxford, Reading, and Newbury, and played a part in the provisioning of the metropolis, it was only below tide-water that it was an efficient artery of commerce.[1] Smaller rivers, such as the Mersey, the Yorkshire Ouse, and the Trent, suffered intermittently from excess or deficiency of water; and the silting up to which they were liable was made worse by the bargemen, who, in order to float their craft over the obstruction, were in the habit of casting their ballast overboard in the shallows.

Under the stimulus of ideas derived from the Dutch, much had been done in the seventeenth century to make the rivers more suitable for transport, and after the Peace of Utrecht the movement was resumed with vigour. In the investment boom of 1717-20 companies were formed to develop the Kennet, Derwent, Douglas, Idle, Mersey and Irwell, and the Weaver—most of them, it will be observed, in the growing coal and textile areas of the north. (London merchants contributed to the finance, but most of the funds came from landowners, traders, and manufacturers who stood to gain by the improved facilities. Of the forty original subscribers to the Mersey and Irwell scheme of 1721, thirty-three belonged to Manchester and three to Liverpool.[2]) By 1724, it is said, some 1,160 miles of English rivers were open to navigation, and, apart from the mountainous areas, most of the country was now within fifteen miles of a navigable waterway.[3]

Improvement took the form not only of removing weirs, strengthening banks, deepening beds, and constructing locks, but also of shortening routes by digging new channels or 'cuts'. From this it was but a short step to the making of continuous artificial waterways, which not only gave speedier transit but, by drawing traffic

---

[1] E. C. R. Hadfield, 'The Thames Navigation and the Canals, 1770-1830', *Ec. Hist. Rev. XIV*, No. 2, p. 172.
[2] A. P. Wadsworth and J. de L. Mann, *op. cit.*, pp. 214-19.
[3] E. C. R. Hadfield, *British Canals, an Illustrated History*, p. 26.

F

away from the old rivers, enabled these to meet more effectively the needs of the fullers and millers for power.

Opinions differ as to whether the Sankey Navigation (initiated in 1755) or the cut from Worsley to Manchester (begun by the second Duke of Bridgewater in 1759) is to be regarded as the first of the English canals.[1] It was certainly the spectacular features of the second of these—the subterranean waterways at Worsley and the aqueduct over the Irwell at Barton—that caught the imagination of the public. The enterprise of the Duke infected men like Lord Gower and Josiah Wedgwood; and the ingenuity of James Brindley was emulated by engineers like Whitworth, Rennie, and Jessop. In the early seventeen-sixties, England entered the canal era.

The motives for making the new navigations were varied. Often the impulse came from landlords eager to extend the market for the corn, timber, and minerals of their estates. It was reinforced by the need of farmers for larger quantities of marl, lime, and manure, and of builders and manufacturers for cheaper supplies of bricks, stone, slate, wool, cotton, and other raw materials. But, above all, the canals came into being because of the rising demand for fuel. The shortage of coal in Liverpool gave rise to the Sankey Navigation. An increasing demand for coal in Manchester led the Duke to dig his canal from Worsley: 'a navigation', he insisted, 'should have coals at the heel of it.' Of the 165 Acts, passed between 1758 and 1802, no fewer than ninety were for concerns whose primary object was to transport coal.[2]

Some canals were constructed by individuals, like the Duke of Bridgewater, Sir Nigel Gresley and the Earl of Thanet.[3] In these cases the capital came from the rents or other income of the proprietor or was borrowed from friends and associates. The Duke mortgaged his lands and obtained loans from relatives (including Lord Gower), from his tenants, from manufacturers (J. and N. Philips of Manchester and John Royds of Rochdale) and from the bank of Messrs. Child & Co.[4] A few short canals were cut by family

[1] Some would set the beginnings of English canals in 1563-5, when a channel was cut to join Exeter to Topsham. The Great Level in the Fens, and the New River from Ware to Islington (the object of which was to bring drinking water to London) may also be regarded as forerunners of the eighteenth-century canals.

[2] J. Phillips, *History of Inland Navigation* (1792), p. 598.

[3] E. C. R. Hadfield, *op. cit.*, pp. 263, 285. Gresley's canal was from Apedale to Newcastle-under-Lyme, the Earl of Thanet's from Skipton to the Leeds and Liverpool navigation.

[4] A. P. Wadsworth and J. de L. Mann, *op. cit.*, pp. 222-3.

businesses, such as those of William Reynolds of Ketley and Samuel Walker of Rotherham.[1] But, in the main, the new navigations were the product of corporate enterprise: between 1758 and 1802, it is estimated, some £13 million was subscribed to joint-stock canals.[2] As with the river companies, the bulk of the money was raised in the locality which the particular project was designed to serve: of the shareholders in the Leeds and Liverpool Canal 71 per cent. lived in Lancashire or Yorkshire.[3]

The first step was to call a general meeting, elect a provisional committee, and appoint an engineer to make a survey and an estimate of cost. Later, a petition was drawn up, and, if all went well, an Act was obtained. Legal and Parliamentary expenses were high, and large sums were sunk in buying off opposition. The cutting of a canal meant interference with private property: owners of the land through which the projected route ran had to be persuaded to relinquish their rights. Sometimes the proprietors of river navigations demanded compensation for loss of tolls: the Duke of Bridgewater was forced to obtain a controlling interest in the Mersey and Irwell Navigation before he could operate his canals.

In the early years of the movement the bulk of the contributions came from well-to-do people who took up shares of relatively high denomination: when the Grand Trunk was floated in 1766 its capital consisted of 580 shares of £200 each. As time went on, however, support was sought from people of less substance. When, in 1792, the Horncastle and Lincoln Canal Company was formed, its shares were of £50; and when, in 1803, the Grand Junction obtained authority to increase its capital (by £400,000), it invited subscriptions for a half, a quarter, and even an eighth, of a share of £100 denomination.[4] Generally payments had to be made at the office of the solicitor to the company; but in the Kennet and Avon Act of 1801 it was provided that some of the new shares should be sold by public auction at the Exchange Coffee House in Bristol or at Garraway's in London. Shares were transferable and were frequently offered for sale by advertisement in local newspapers. It was usual for a company to pay interest, at the legal rate, from the date of

---

[1] *Minutes relating to Samuel Walker & Co., Rotherham* (ed. A. H. John). Coun. for the Preservation of Business Archives (1951), p. 5.
[2] J. Phillips, *op. cit.*, p. 598.
[3] In 1772 the University of Glasgow subscribed £200 to the Monkland Canal. H. W. Dickinson and Rhys Jenkins, *James Watt and the Steam Engine*, p. 32.
[4] J. Phillips, *op. cit.*, pp. 126, 280, 311.

subscription to that of completion of the enterprise, after which the dividend depended on profits. Since it was often several years, or even decades, before the work could be finished, the burden of fixed charges was heavy, and it was sometimes necessary to obtain power from Parliament to reduce the rate of interest paid to shareholders.[1] Frequently companies were given permission to raise additional funds on mortgage or by the issue of bonds or promissory notes. Under the 47 Acts relating to inland navigations passed in the years 1788-95, over £6 million was to be raised by subscription, and nearly £2 million by borrowing. The canal companies did much to popularise not merely the equity, but also the mortgage, the bond, and the debenture.[2]

Much has been written about the engineers who surmounted the physical barriers to an efficient system of waterways. Less is known about the men who put their plans into effect. Beside the consultant engineer there was usually a clerk of the works, or resident engineer, who had to procure supplies of material and equipment, assemble large numbers of labourers (often at places remote from a town), make arrangements for their board and lodging, and supervise the execution of the work.[3] James Brindley, who combined the functions of surveyor and resident engineer, has been acclaimed for his technical skill and his facility for improvisation: it seems likely that (as in the case of Richard Arkwright) the real secret of his success lay in his gift for organisation. He had had experience as a master millwright and had, no doubt, learnt how to lead and discipline workers.[4] But to control 400 to 600 rough and turbulent men scattered over miles of country was a more difficult problem: his solution, and that of his successors, was found in devolution. Work of a special nature was put out to contractors, who were paid so much for each lock or bridge constructed. But for excavating the main sections of the canal it was usual to enter into agreements with groups of labourers, at a rate of 3*d*. (or some similar sum) for each cubic yard of earth

---

[1] Under an Act of 1769 the Oxford Canal had to pay 5 per cent. to the subscribers. When, in 1775, powers were obtained to borrow £70,000 on mortgage 'at legal or less interest' the percentage payable to the original subscribers was reduced to 4. An Act of 1786 provided that this 4 per cent. should be a maximum, and not an obligatory rate, until £30,000 of the loan should have been repaid. Information provided by Dr. L. S. Pressnell.

[2] *Reflections on the Present State of the Resources of the Country* (1796), p. 14.

[3] E. C. R. Hadfield, *op. cit.*, pp. 39-40.

[4] A. P. Wadsworth and J. de L. Mann, *op. cit.*, p. 305.

or stone removed.[1] The money earned was handed over periodically in a lump sum to the leader or spokesman of the gang, and the distribution was probably left for the men themselves to determine. The following entry from a book of accounts of the Duke of Bridgewater suggests that, in 1771, a labourer might expect to earn a little over 5s. a week, though whether this was with, or without, board and lodging is not known.

June 13, 1771. By p^d Edward Topping & 32 more
Labourers for a month's wages ending May 3,
Cutting the canal at Runcorn.   £35 14s 6½d.

Apart from the labourers, who worked with pick and shovel, large numbers of brick-makers, quarrymen, bricklayers, and miners or tunnellers were employed.[2] There was also a need for skilled artisans and for more elaborate tools and equipment: the Duke of Bridgewater had barges fitted out as mason's or carpenter's or blacksmith's shops, so that craftsmen could be on the spot as the work advanced.

Of the geographical and social origins of the men who made the canals little is known. Some, it is said, came from the Fen country, others from Cumberland, Northumberland, Scotland, Wales, and Ireland; but it seems likely that the majority of those who dug the early 'cuts' of the north were hillmen from the Pennines. Certain it is that the movement of large numbers of vigorous men about the countryside had important effects on English society. It must have brought new ideas to rustic villagers and increased the mobility of labour as a whole. But, most important of all, was the creation of the navvy himself—that unique type of Englishman, who was to rise to international fame, through his work on the railways of more than one continent in the middle decades of the nineteenth century.

[1] The estimates of Robert Whitworth for the Grand Trunk and of John Smeaton for the Forth and Clyde are based on a rate of 3d. a cubic yard. J. Phillips, *op. cit.*, pp. 160-4, 512, 536.

[2] The following entries from the records of the Duke of Bridgewater indicate that brickmaking was done by contract:
    1769, Oct. 17, Alex. Bradley, Brickmaker £130 13 0, the amount of his Bill for making 455 Thousand of Bricks at the Water Meetings.
    Oct. 30, James Partington, For making 93 Thousand of Bricks for his Grace's Use £25 5 6.

IV

The reform of the highways was no less badly needed than that of the waterways. Some of the roads were described by a pamphleteer as being 'what God left them after the Flood'. The roads of many parts of rural England, no doubt, sufficed for the farmer, the local carrier, and the pedlar; and the stone causeways laid in the middle or by the side of the tracks in the north, served well enough for the pack-horse with its burden of woollens or linens. It was the main roads connecting the metropolis with the provinces that gave rise to most complaints from travellers and traders. Foundations were often weak, and drainage was neglected. The narrow wheels of carts and wagons made ruts in the loosely packed surface material, and the hooves of cattle helped to churn the roads into bogs. At times some of the highways through the clays of the midlands became impassable to heavily loaded vehicles, so that commodities in the cost of which transport charges played a large part rose steeply in price. In May, 1751, as a result of heavy rains, the price of coal rose at Derby from 4*d*. to 8*d*., at Rugby from 8*d*. to 14*d*., and at Northampton from 10*d*. to 18*d*. per cwt.[1]

Generally, movement by road, other than on horseback, was painfully slow. When, in 1739, Roderick Random decided to leave Newcastle for London, it was natural that he should first look for a passage by ship. His friend, Strap, however, warned him of the perils of a coastwise voyage in the winter and persuaded him to make his way by land. The fact that, travelling on foot, he was able to overtake the wagon that had set out two days earlier, may serve as evidence of the pace of wheeled transport in the days of Tobias Smollett.

Part of the trouble arose from the fact that the administrative body responsible was unfitted to its task. Under a statute of 1555 the duty of caring for the highways had been put on the parish. Each parish appointed surveyors—unpaid and unskilled—to supervise the labour of the parishioners who were called on to contribute the services of horses and carts, or to give four or six days a year to the repair of the roads in their own area. During the Interregnum the surveyors had been given power to levy assessments and to hire workers to supplement the unpaid statute labour. But though, in some parts of the country, paupers were put to work on the roads,

[1] *Gentleman's Magazine*, May, 1751.

few of those employed had any special qualifications for the work, and in parishes where through traffic was growing the condition of the highways probably deteriorated.

The measures taken by Parliament to arrest the decay took the form of defining the kinds of traffic to be allowed on the roads. Restrictions were set on the number of horses to a wagon and on the load to be carried in any single vehicle. Throughout the century there was much controversy as to the relative merits of broad and narrow wheels. It was generally held that the wider the rim of the wheel the better for the road, but the worse for the vehicle. The well-to-do wanted narrow wheels on the coaches, in the interests of speed. They were favourable, however, to legislation imposing the use of broad wheels on wagons, not only because these would do less damage to the road surface, but because if the rut made by the wagon wheel were sufficiently broad it might serve as a riding track for the gentleman on horseback. The Highways Act of 1753 prohibited the use of wagons or carts (but not coaches) with fellies of less than nine inches across. But much ingenuity was shown in circumventing the regulation; in the following year, therefore, narrow wheels were permitted, but wagons with broad wheels were allowed to pass free or at lower rates of toll; and under the General Turnpike Act of 1773 an elaborate system of differential tolls was established. Vehicles were classified according to the breadth of their wheels: the rates fell as the breadth rose, and wagons with 'rollers' of between thirteen and sixteen inches in width were exempted from tolls for a year and given preferential treatment thereafter.[1] But, again, the wheelwrights succeeded in outwitting the legislators. Wagons were fitted with broad rims, but these were rounded or made of three separate bands of iron, the middle one of which projected, so that, in either case, only a small section of the felly came into contact with hard ground. What was ostensibly a broad wheel had the same effect on the road surface as the narrow wheel that subjected the wagoner to a high toll. But long before 1773 it had become evident that reformers must direct their attention less to the traffic and more to the foundations and surface of the road.

Much of the work of improvement has left little record. Often when the land of a parish was enclosed the opportunity was taken

---

[1] Arthur Cossons, 'The Turnpike Roads of Nottinghamshire', *Historical Association Leaflet*, No. 97 (1934), pp. 20-2.

to widen or straighten an old road or to make a new one across what had previously been open arable fields or pastures. Proprietors of coal-mines, iron-works, and other industrial enterprises did much for the ways along which their raw materials or products were carried: Samuel Walker's Diary for November 1758 records[1] that he had made 'almost incredible improvements in the road to the Holmes from Masbro' and Kimberworth, and in ye Lanes towards Tinsley'. But generally the task of reforming the highways was assumed by corporate bodies created for the purpose. As early as 1663 powers had been given to the Justices of the counties of Hertford, Cambridge, and Huntingdon to set up turnpikes and collect tolls on vehicles and livestock passing along the stretch of the Great North Road under their control.[2] It was not until early in the eighteenth century, however, that the principle of vesting the administration of roads in *ad hoc* local bodies, and of transferring the cost of maintenance from the public to the users, was firmly established. One of the first turnpike trusts was formed in 1706-7 to improve the section of the London-Holyhead highway between Fonthill and Stony Stratford. Its success led to the passing of hundreds of Acts extending the system to almost all parts of England. The trustees, who were generally landowners, merchants, or industrialists, were given powers to raise capital, usually at 4 or 5 per cent., and to apply this to making a new road (or, more often, to improving an old one) in a particular locality. The franchise was for a defined period, usually twenty-one years, but the term was often extended by further legislation. At each end of the road, gates were erected at which tolls were collected to cover the cost of maintenance, and (though the aspiration was rarely achieved) to repay the capital subscribed. The trust was often given the right to the statute labour of the area concerned, or was allowed to accept money, from individuals or the parishes, in lieu of this. In the main, however, it employed local labour, at rates of pay which (if we may accept Dr. Gilboy's figures for the North Riding as typical) rose from about 12d. a day in the 'fifties and 14d. in the mid-sixties, to 19d. in the 'seventies, and to 20d. in 1786. The trust appointed its own officers (an honorary treasurer, a surveyor, and a secretary), as well as gate-keepers and collectors, though it was a common practice to farm out the tolls, for a year or more at a time, to individuals selected by competitive bids at an auction.

[1] *Op. cit.*, p. 5.   [2] A. Cossons, *op. cit.*

The employment of professional surveyors and road-makers effected some change. John Metcalf, the blind engineer of Knaresborough, made about 180 miles of turnpike road in the north. His method was to dig out the soft soil, lay bunches of ling and heather on the bed of soft earth, cover these with stone, and dress with a convex layer of gravel, so that the rain water ran off to the ditches made at each side. But there were many others who, each adapting his skill to local variations of sub-soil and local needs, opened up roads in all parts of the kingdom. (Perhaps the most important change was the substitution of a convex for a flat surface.) The result was a general quickening of trade. 'As to new markets at home' wrote Dean Tucker,[1] 'every road well made has that effect in one degree or other.' And, as with the canals, the main benefit came from the cheaper and more rapid transport of heavy materials, and, most of all, that of coal. After the boom in the formation of trusts in the early 'nineties, in which Macadam and Telford rose to fame, England had a network of roads not badly adapted to her requirements and far superior to that in most other parts of the world.

V

In the eighteenth century no sharp line was drawn between what it is now the fashion to call the public and private sectors of the economy. The unreformed municipal corporations have been charged with indolence and corruption, but some of them showed initiative not only in paving and lighting the streets, making sewers, and erecting public buildings, but also developing communications outside their administrative boundaries. The City of London played a part in building bridges across the Thames and providing facilities for shipping. The corporation of Liverpool constructed docks, improved the navigation of the Mersey and the Weaver, took part in initiating the Sankey Navigation, and made plans for artificial waterways to connect the town with Hull and other ports.[2] The Corporation of Bath took the responsibility for improving the River Avon. And the council of Doncaster shared for a time with the Cutlers' Company of Hallamshire the task of improving the River Don.[3] Since there was a tendency to identify the interests of the community with those of the owners of property, well-to-do citizens often shared in

[1] J. Tucker, *Instructions to Travellers*, p. 13.
[2] T. S. Willan in *Ec. Hist. Rev.* vol. *VIII*, No. 1.
[3] T. S. Ashton, *Iron and Steel in the Industrial Revolution*, p. 245.

such public enterprises. In 1771 all people assessed at £100 for Land Tax in seven counties were included among the commissioners for the control of the Thames,[1] and, under the Manchester Improvement Act of 1776 and the Police Act of 1792, townspeople who subscribed a minimum amount automatically became commissioners for carrying out these measures.[2]

The central government also took a share in the movement for better communications. Consideration of strategy led it to drive roads through the Scottish Highlands after 1745, to make loans to the Forth and Clyde, and the Caledonian Canal Companies in 1784 and 1799, and to construct military canals in the later phase of the struggle with Napoleon. Parliament occasionally made contributions to works of public utility, as when, in 1757, it subscribed £15,000 to the re-building of London Bridge. But generally its functions were to regulate rather than initiate. Not the least important of its contributions to the development of communications were those made by the Acts of 1773 which provided that milestones should be set up on the highways, signposts erected at road junctions, and that bridges should be walled or fenced.[3] In conferring powers on canal companies and turnpike trusts it laid down conditions as to membership and policy. It sometimes stipulated that Justices of the Peace should serve, *ex officio*, on a turnpike trust; and, in order to prevent undue concentration of power, it set a limit to the number of shares that might be held by any one person in a road or canal undertaking.[4] The Duke of Bridgewater was both a canal owner and a carrier: he was forbidden to charge freight-rates of more than 2s. 6d. a ton, or to sell his coals in Manchester and Salford at a price above 4d. a hundredweight. But neither the turnpike trusts nor the joint-stock canals had the right to act as carriers, so that, in their case, regulation related only to tolls. Generally they were required to allow lime and manure for the land, and stone and gravel for the repair of the roads, to pass free of charge; and the same exemption was given to troops and their equipment. Travellers on foot, the carriages of householders on their way to church or to the hustings, and the carts of husbandmen whose farms

---

[1] E. C. R. Hadfield, *loc. cit.*
[2] A. Redford, *The History of Local Government in Manchester*, vol. I, pp. 157-8.
[3] D. Macpherson, *op. cit.*, vol. III, p. 547.
[4] For example, no individual was allowed to hold more than twenty shares (of £200 denomination) in the Grand Junction canal in 1766.

Internal Trade and Transport

adjoined the highway, also went free of charge. Rates of toll varied with the type of vehicle, the number of horses, the weight of the load, and the value of the commodity carried. Some trusts differentiated between summer and winter traffic, or day and night traffic, and occasionally higher tolls were imposed on Sundays. Local interests secured preferential treatment for coal on the turnpikes in the county of Nottingham; and in 1793 the Derby Canal Company was required to allow 5,000 tons of coal annually to go without toll for the use of the poor of Derby.

Many of the Acts authorising the formation of canals or turnpikes set limits to profits. In the case of the trusts dividends were usually restricted to 5 per cent. Any additional profits were to go to the repayment of mortgages or the reduction or removal of tolls. In 1697-8 the proprietors of the River Tone Navigation agreed to hand over all profits above 7 per cent. for the relief of the poor at Taunton. In 1721 the Act for the Weaver Navigation provided that, after 6 per cent. had been paid to the subscribers, any balance of income should go to the upkeep of highways and bridges. And in Acts for the Derby Canal and the Hornsey and Lincoln Canal it was laid down that if profits exceeded 8 per cent. tolls should be lowered. To a large extent, the subscribers to both canals and turnpikes had their reward less in dividends than in the enhancement of value of their estates or the earning power of their business concerns. But the unpaid Treasurer of the turnpike had the right to employ accumulated funds for his own purposes until they were needed, and no doubt others concerned with canals had similar perquisites. Generally administration seems to have been honest and reasonably efficient. The traditions of the modern public utility owe much to the canal companies and the turnpike trusts.

VI

The construction of the new means of transport absorbed large resources. It was not simply a matter of dredging a river, resurfacing a highway, or digging the bed of a canal: the roads had to be supplied with fences, gates, and toll houses, and the waterways with an elaborate system of reservoirs, locks, sluices, towpaths, bridges, and quays. New types of vehicles and craft were needed, as well as new inns, stables, and warehouses. Subsidiary channels had to be cut, or railroads laid down, to connect works or coalmines with the canal or turnpike, and for these also new tugboats or special types of rolling stock were required.

It was not easy to make accurate forecasts of the return on such investment. Much depended on the costs of construction and maintenance, and high among these was the cost of money. As borrowers, the companies and trusts were subject to competition. When the government was living within its income they could usually obtain funds at, or below, the legal rate of interest: when it was drawing heavily on the market they had difficulty in getting money at all. For this reason, among others, the improvement of transport was not a continuous process: it was marked by sudden spurts of energy, each followed by quiescence or stagnation.

There was liveliness in the development of rivers in 1697-1700 and 1718-20—both periods of relatively low rates of interest. Numerous turnpike Acts were passed in 1723-5, 1729-33, 1739-43, 1750-3, and 1757-62. Large numbers of projects for canals, as well as turnpikes, were put forward in the periods of cheap money, 1764-9, 1772-7, and 1787-94; and at the very end of the century, in 1800, a new canal boom was well on the way. It must not be thought that the intervening years saw no development, for work on a project might continue for months or even years after a boom was over. But when, as in 1777-87, the break was a long one, it is safe to assume a decline of construction. The discontinuity had widespread effects. The navvies and labourers were engaged in work which, whatever its future return, added nothing immediately to the supply of goods for consumption. Their demand for food, drink, clothing, and shelter created employment for others: high activity in construction led to prosperity in a variety of occupations. When recession occurred the consequences for employment were similarly wide. If it came in the middle of a war, as in 1778, those displaced might find occupation in the armed forces or in industries supplying munitions. But in years such as 1754, 1763, 1774, and 1783, no such compensation was afforded.

The benefits expected from the use (as distinct from the construction) of the new means of transport were varied. Projectors pitched their hopes high. It was claimed that improved roads and waterways would bring about a fall in the prices of foodstuffs, and put an end to the threat of famine. They would reduce the number of horses, and so free resources for the production of wheat. By lowering the cost of lime and manure, they would raise the quality, and hence enhance the value, of land. They would encourage the mining of coal and other minerals, give rise to new industries, extend the trade of the ports, and reduce the number of middlemen. They would create new demands for labour and check the emigration of skilled workers. And they would greatly increase the mobility of troops and artillery in times of war or civil disorder.[1]

Not all these results appeared immediately, and some were never fully realised. For this, vested interests and narrow views were partly responsible. It was not only the proprietors of the river navigations who put obstacles in the way of the projectors of turnpikes and canals; landlords, farmers, and manufacturers whose properties were at a distance from a proposed new route feared competition from those able to make use of it. More than that: they might be put to increased expense in maintaining the old means of communication. When, for example, in 1771, it was proposed to make a canal from Reading to Monkey Island, landowners about Henley protested that this would injure their interests as suppliers of timber and firewood to London. The canal would not pass near their own estates, and the charge of maintaining the old river would bear heavily when the boats of more favourably situated proprietors had been deflected to the proposed navigation.[2] Carriers, and traders in river ports, such as Bewdley and Bridgenorth, were no less obstructive, as were also the innkeepers on the traditional routes.

The canals and turnpikes represented a substitution of capital for labour. Where the old and new means of transport operated side by side, this fact could hardly fail to impress the observer. At Barton Bridge, John Phillips[3] pointed out, one could see at the same time 'seven or eight stout fellows labouring like slaves to drag a boat slowly up the Irwell, and one horse or mule, or sometimes two men at the most, drawing five or six of the Duke's barges, linked together, at a great rate upon the canal'. To many publicists of the eighteenth century anything that seemed likely to reduce employment was objectionable. It was feared that the canals would be the ruin of the carriers by road. Ardent advocate, though he was, of inland navigation, Richard Whitworth[4] declared himself 'unwilling to unbend the crooked finger, or straighten the almost distorted joint,

---

[1] For these and other claims see James Sharp, *Survey of the Canal from Waltham Abbey* . . . (1733); John Phillips, *Treatise on Inland Navigation* (1785); Joseph Priestley, *Historical Account of Navigable Rivers, Canals, and Railways* (1830).
[2] *J.H.C.*, vol. xxx (1771), p. 83.   [3] J. Phillips, *op. cit.*, p. 87.
[4] R. Whitworth, *The Advantages of Inland Navigation* (1766), p. 29.

inured to tally with the stroke of its accustomed trade'; and in the interests of the wagoners he urged that no canal should be allowed to pass within four miles of any large manufacturing or trading town.

Fears were expressed lest the turnpikes should make it possible for farmers at a distance to compete with those close to London and that agricultural rents in the home counties would fall. Here, however, it is necessary to distinguish between various agricultural products. Even under the old forms of transport there had been little difficulty in carrying wheat over long distances, and bread seems to have been as cheap in London as in many agricultural counties. The same was true of beef and mutton, for the unreformed roads served well enough for cattle and sheep on their way to the market. But, as Arthur Young pointed out, the prices of butter, veal, mutton, and other perishable goods tended to be high in the metropolis and to vary in other places inversely with the distance from this.[1] Improvements in transport would reduce the differentials, with adverse effects on dairy farmers in the areas near to the capital. They would not only lower prices in the central market, but would raise them in the agricultural provinces. 'All the sensible people', wrote Young in 1769, 'attributed the dearness of their country to the turnpike roads; and reason speaks the truth of their opinion . . . make but a turnpike road through their country and all the cheapness vanishes at once.' Agreeable as this might be to the producers in these areas, it was naturally resented by the consumers. In times of dearth mobs not infrequently threw down turnpike gates and destroyed toll houses, partly, no doubt, because these impeded the flow of food from outside, but also because they belonged to an agency that carried away local supplies to the cities.

The claim that the canals would increase national security was countered by the argument that they would reduce the number of horses. At the height of the transport boom in 1767 Charles Homer declared that 'the carriage of grain, coal, merchandise, etc., is in general conducted with little more than half the number of horses with which it formerly was'. This, it was urged, might have serious effects on the army in time of war. More important, the canals and turnpikes would draw trade from the coastal vessels and reduce the number of sailors available for service with the fleet. In view of the fact that the men in the coasters were renowned for their physique

---

[1] Arthur Young, *Tour through the Southern Counties of England* (second ed., 1769), pp. 312, 315.

and hardihood, the reply to this charge can hardly have carried much weight. It was that the inland waterways would themselves breed mariners, since they would offer experience in the handling of ropes and rudders, and 'since every boy in each village through which the canals run, will have a desire to become a seaman'.[1] There was, however, perhaps more force in the subsidiary claim that the inland navigations might offer employment to men crippled in His Majesty's service.[2]

In the event, most of the fears proved groundless. Some of the old river ports fell into decay, but compensation was found in the rise of new transhipment centres such as Stourport and Coalport. More men, not fewer, found employment in transport: the saving of labour per ton carried was more than outweighed by the increased tonnage that resulted from the lowering of freights. There was no decline in the coasting trade, but an increase—brought about by the greater volume of goods that reached the ports from the inland centres. Nor was there any reduction in the number of horses. The expansion of traffic on both roads and canals was such that more were required, and in 1773 John Arbuthnot could assert that over the past twenty years the number of post-horses had increased tenfold—at a time when horses were finding increasing employment not only in the army and the hunting field, but also in and about collieries and other industrial establishments.

On the other hand, not all the hopes of the innovators were realised. The turnpike trusts were relatively small undertakings, each concerned with a few miles of road.[3] It was expedient to attract not only local, but also through traffic, and hence the trustees of each turnpike sought to link up their road with others. By 1765 there was a line of turnpikes from London to Berwick-on-Tweed, with only a short break in the neighbourhood of Doncaster;[4] and similar chains were either in being or in course of construction to Falmouth, Bristol, Carlisle, and other remote places. Since, however, tolls had to be paid every seven or ten miles, long-distance traffic consisted largely of passengers and of goods of small bulk and high value: it was only over relatively short distances that use

---

[1] R. Whitworth, *op. cit.*, p. 35.
[2] J. Phillips, *op. cit.*, p. xi. In 1774 it was estimated that the navigation at Liverpool might supply a thousand men for the navy. Lieut. Tomlinson, *loc. cit.*, p. 15.
[3] Ten to twenty miles was normal.
[4] Arthur Cossons, *op. cit.*, p. 12.

was made of the turnpikes to transport heavy commodities. Nevertheless, even a limited widening of the market was important. Farms benefited by the greater ease of obtaining lime and manures, and towns by ready access to sources of fuel and building materials. In the area within a radius of forty miles from London, and in the heavy clay lands of the south and east midlands, the trusts brought great gains to both producers and consumers.

The navigation companies were regional rather than local organisations: each was formed to meet the needs of a fairly large area. The object of the Bridgewater canal (1759-61) was to provide a substitute for the Mersey and Irwell navigation with its winding course, currents, shallows, and its liability to droughts and floods. The canal linked up effectively south-east Lancashire with the sea and played an important part in the rise of both Manchester and Liverpool. The Trent and Mersey canal (1766-77) offered an alternative to the inconvenient Weaver navigation: it provided the salt-makers of Cheshire and the pottery manufacturers of Staffordshire with cheap transport to the mouth of the Mersey. In the same way the Staffordshire and Worcestershire and the Birmingham canals (1766-72 and 1768-9) brought to the varied industries of the west midlands direct access to the artery of the Severn. It would be tedious to continue the list. The distances over which bulky goods were carried were greater than in the case of the turnpikes, and the policy adopted by the companies tended to this end. Costs of carriage were made up of tolls and haulage charges. The Duke of Bridgewater was unique in that he was both a canal proprietor and a carrier. The Mersey and Irwell navigation had charged 40s. a ton for transporting goods from Manchester to Liverpool: the Duke charged less than one-fifth of this. Rates between intermediate points varied with the distance, but not in strict proportion: the cost per ton-mile fell as the mileage increased.[1] In the same way the Grand Junction (and no doubt other companies) levied tolls which increased with the distance, but only up to a certain point: beyond this, goods went toll-free.[2] The arrangement did not apply to haulage charges, but since tolls were usually a larger element in costs than these, the stimulus

---

[1] In 1794 the charge for carrying goods between Manchester and Liverpool by canal was 7s. 2d. a ton. The charges from Manchester to intermediate points on the route were as follows: Broadheath, 3s. 8d.; Stockton, 5s. 2d.; Preston, 6s. 2d.; Runcorn, 6s. 2d. Higher freights were charged for cottons: to Stockton, 6s. 8d.; to Preston, 8s.; and to Liverpool, 10s. 6d. *Scholes's Manchester and Salford Directory*, 1794.

[2] *Report on the Coal Trade* (1800), p. 645.

to long-distance traffic was significant. In 1792 the cost of sending a ton of goods from Birmingham to Liverpool was £5 by land and £1 10s. by water: from Birmingham to Gainsborough the corresponding charges were £3 18s. and £1.[1] It was for transport over such distances that the saving was most marked.

As has already been said, most of the early canal companies had as their prime object the linking of an industrial area with the sea. But the leading engineers had in their minds a grand plan to connect, by a series of navigations, the chief ports of the country. The scheme for the 'Cross' went well for a time: by 1772 Bristol was joined to Hull, and by 1777 Liverpool was connected by inland water with both these places. But thirteen more years were to pass before London was brought into the system. Many circumstances accounted for the delay. The porous subsoil of the London area presented technical difficulties. The city had already adequate connections, by river and sea, with its markets at home and abroad: there was no such urgency as in the case of Manchester and Birmingham. After 1777 rates of interest rose sharply and remained high till towards the end of the 'eighties: the time was unpropitious for large enterprises. In 1790, however, with the completion of the Coventry and Oxford canals, the capital had access by inland waterways to Bristol, Liverpool, and Hull; and a new route to the west was made by the Thames and Severn navigation.

The Cross brought less gain than some had expected. It consisted of several separate waterways each of which claimed its toll. A trader who wished to send goods from an outport to London, or from one outport to another, usually found it better to do so by sea. And lega-obstructions prevented London from reaping the full benefits of inland navigation. Partly because of concern for the 'nursery of sea men', partly because the City of London obtained substantial revenue

[1] *Felix Farley's Bristol Journal*, December 1st, 1792. Other instances given here are:

	By land £ s.	By canal £ s.
Manchester to Birmingham	4 0	1 10
,, ,, Stourport	4 13	1 10
,, ,, the Potteries	2 15	15
Liverpool to Wolverhampton	5 0	1 5
,, ,, Stourport	5 0	1 10
Chester to Wolverhampton	3 10	1 15
,, ,, Birmingham	3 10	2 0

In some cases there would have been a cost in carrying the goods from the point of origin to the canal.

from the tax on sea-borne coal, it was deemed impolitic to allow minerals to be transported by inland navigation beyond points some miles distant from the capital. And even after 1800, when the barriers were lifted, coal carried in this way, like that carried by sea, was subject to high duties. The canals brought no such benefits to London as the railways were to bring in the nineteenth century. They were a northern innovation, and their chief effect was to be seen in the rapid rise of industry in Lancashire, the West Riding, and the Midlands, rather than in any large change in the metropolitan area.

After the Cross had been completed enterprise and capital turned to less ambitious schemes. One or two of the later canals, such as the Kennet and Avon and the Liverpool and Leeds, were on a large scale; but most of the navigations of the 'nineties had the object of joining a coal field, an iron centre, or a manufacturing area with a nearby port or an existing trunk line of waterways. In South Wales, in particular, the result on the growth of population and industry was spectacular.

Stress has been laid on the magnitude of the investment in turnpikes and canals, and on the labour-saving effect of these devices. In the long run, however, the economy of capital was no less pronounced. Before the improvements it was not uncommon for traffic to be held up for weeks by bad weather, and for this reason the pithead price of coal was sometimes lower in winter than in summer.[1] Ice or flood sometimes delayed traffic on both turnpikes and navigations, and there were times when the canal companies had to refuse cargoes because of drought. But, generally, the new forms of transport were less affected by adverse conditions than the rivers and unreformed roads. They provided a quicker service, and so, at any point of time, a smaller proportion of the wealth of the nation consisted of materials and finished goods in transit. Hence more resources were available for investment in fixed form. The growth of factories and power-driven machines in the late eighteenth century was the result not only of the opening of new markets, but also of the capital-saving effects of the innovations in transport.

Report on the Coal Trade (1800), p. 644.

CHAPTER FOUR

Manufactures

I

IN 1700 most Englishmen were engaged in primary production: they were husbandmen, graziers, shepherds, fishermen, miners, and quarrymen. Most of their needs, other than those supplied by their own efforts, were met by secondary producers close at hand: each country town had its miller, maltster, brewer, tanner, and sawyer, each village its baker, smith, and cobbler. These rural craftsmen must have outnumbered the workers in several of the more highly organised industries. But they have left few records, and hence little can be said of them here.

There were areas where natural conditions or historical circumstance had led to a specialisation of activities. Here and there, on the coast, communities of fishermen and sailors had grown into trading centres, and in some of these commerce had given rise to manufactures. Few ports confined themselves to a single industry. London, in particular, had so many roles to play that no one activity could dominate. As the chief resort of merchant vessels, the Thames estuary had drawn to itself shipbuilders, anchor-smiths, ropemakers, coopers, and sailcloth manufacturers; and the city was the home of craftsmen who made sextants, chronometers, telescopes, and other things necessary to ships. Since Londoners handled the bulk of the raw material brought from abroad, it was natural that they should concern themselves with its fabrication: hence, for example, arose the throwing and weaving of silk in Spitalfields. Since London merchants were responsible for the quality and appearance of English exports, it was natural, again, that they should wish to control the later stages of production; and so processes like calico-printing were centred on London. Most immigrants from Europe came first to the capital, and industries like paper-making, based on their skills, settled in the metropolitan area. Mark Lane and Smithfield provided material for corn-mills, breweries, distilleries, and tanneries, as well as for works making soap, candles, glue, starch, and other by-products of cereals and livestock. Finally, as the centre of taste and fashion, London took the leading part in the manufacture of consumers' goods of high quality, including coaches, furniture, millinery, silverware, and jewellery.

Much the same, though in less degree, was true of the other ports. Bristol built ships, worked up West Indian products such as sugar and tobacco, and made glass, pottery, and brass from materials brought by river or sea. Newcastle was concerned primarily with dealings in coal, but she had shipyards, steel furnaces, and establishments for making anchors, chains, picks and shovels, and cutlery. Liverpool built ships, manufactured tools and watches, and played a large part in the development of the salt-works of Cheshire and the glass industry of south-west Lancashire.

Geological conditions obviously determined the location of mining. Tin was still worked in Cornwall; but before 1700 the miners had begun to penetrate below the tin-bearing rock, and by the middle of the century copper had become the mainstay of the Cornish economy. In 1768, when the ores of Parys Mountain were first exploited, another large community of copper-miners came into being in Anglesey. In the Mendips supplies of lead were virtually exhausted, but lead-mining continued to flourish in Cumberland, West Durham, and Derbyshire. The centre of the salt industry had moved from the north-east coast to Cheshire, where rock salt, as well as brine, was now worked. Clays for pottery were excavated in Staffordshire, and increasingly in Cornwall; and the Stourbridge area provided material for crucibles, furnace-linings, and fire-resistant bricks. Cumberland and North Wales had slate quarries, and Portland was famous for its stone. By far the most important geological influence, however, was exercised by the distribution of coal. When this was found in conjunction with other minerals there was a strong tendency to the localisation of manufacture. After 1709, when Abraham Darby succeeded in smelting ore with coke, the iron industry concentrated in the colliery areas of Shropshire and Worcestershire, and, later, in Staffordshire, Yorkshire, and South Wales. Coal and iron in association had already led to an assemblage of the hardware trades in the Black Country, and of tool-making in south-west Lancashire; and access to coal, as well as supplies of steel and gritstone, accounted for a concentration of the cutlery trades in and about Sheffield. Long before coal was used as a source of industrial power it was important to the manufacture of textiles, for the water needed for cleansing the wool and fulling and dyeing the cloth had to be heated. Local supplies of coal, as much as any other factor, account for the concentration of the woollen industry in the West Riding, and the fustian and cotton manufactures in Lancashire.

Coal could be carried for industrial, as well as domestic, use to distant places. Without the supplies brought from the Tyne and Wear, Norwich, and even London itself, could hardly have built up large manufactures. South Wales provided fuel for the engines that pumped water from the mines in Cornwall. Coal brought by the Severn from Shropshire aided the development of the sugar and glass manufactures of Bristol. And supplies from South Lancashire were transported across the Mersey and up the Weaver to salt-works at Northwich, Middlewich, and Nantwich.[1] In each case, however, the costs were high. Generally the ships or flats had to return in ballast, and for this reason freights on the voyage back to the coal-field were much lower than those on the outward trip. Ambrose Crowley selected Sunderland as the site for his works for this reason: it was more economical to send iron to the north than to bring coal to the forges in the Weald.[2] Cornishmen complained of the absence of manufactures in their country.[3] But it was obviously better for them to confine their activities to mining and the preliminary process of crushing the ore than to suffer the costs of transporting heavy and bulky fuel from Newcastle or South Wales. Hence copper smelting developed at places like Neath and Swansea, and tinplate manufacture at Pontypool and Kidwelly, rather than at Falmouth or Truro. And hence also pottery was made of Cornish clay, not in Cornwall, but in distant north Staffordshire.

Perhaps, however, the most powerful influence of all on localisation was that exercised by water. Large volumes of water were required by the paper-makers who settled along the valleys of the home counties and Hampshire, and of the tanners who congregated about London, south of the river. Water was needed by the linen industry for the retting of flax and, as mentioned above, by the woollen and cotton industries for a variety of purposes. The textile trades abhorred hard water. And so, though they existed in almost all parts of England, they congregated in East Anglia and about

---

[1] For the developments of transport that made this possible see T. C. Barker, 'Lancashire Coal, Cheshire Salt, and the Rise of Liverpool', *Transactions of the Historical Society of Lancashire and Cheshire*, vol. 103 (1951).

[2] M. W. Flinn, 'Sir Ambrose Crowley, Ironmonger, 1658-1713', *Explorations in Entrepreneurial History*, vol. V, No. 3 (1953), pp. 164-5.

[3] John Rowe, *Cornwall in the Age of the Industrial Revolution*, pp. 19-20.

Dartmoor, the Cotswolds, and the Pennines, where the streams were pure and free from lime. Many industries made use of water for power. Flowing water was employed to work the pumps of the miners, the machines of the paper-makers, the fulling mills of the woollen manufacturers, the bellows, hammers, and slitting mills of the iron masters, and the grindstones of the Sheffield cutlers and Black Country tool-makers. (Wolverhampton, it is said, turned to the manufacture of locks and bolts because lack of water precluded the making of edge tools.) Much of the capital investment of the eighteenth century went into mill ponds, conduits, and water-wheels.[1] In areas where water could be obtained only at high cost there could be little development of manufacture.

It is sometimes asserted that industries were attracted to areas where the cost of living was low. An additional reason given for the setting-up by Ambrose Crowley of works at Sunderland is the low price of food in the area,[2] and a recent writer claims that it was the cheapness of victuals, among other influences, that had led to the growth of the iron industry in Crowley's native county of Worcester.[3] The implication is that cheap food meant low labour costs—in spite of Adam Smith's view to the contrary. It is beyond doubt that employers often transferred their activities from corporate towns in order to escape from restrictions imposed by privileged groups of workers, or from municipal or governmental regulations as to labour. But the movement of industry was rarely induced by the prospect of lower wages in the new area. Quite the reverse: 'we daily see manufacturers leaving the places where wages are low and removing to others where they are high', it was said[4] in 1752. And even if, now and then, an employer sought out a place where he could obtain workers at small rates of pay, as others followed suit the rising demand for labour and food must soon have put an end to his hopes.

In 1700 England was already parcelled out into industrial provinces. In each of these there was usually some specialisation by

---

[1] In 1770 Sheffield had no fewer than 133 wheels, and 896 separate troughs for grinding cutlery, in addition to 96 water-driven forges, tilts, corn mills, and other establishments. G. I. H. Lloyd, *The Cutlery Trades*, pp. 157, 443.

[2] M. W. Flinn, *loc. cit.*, p. 165.   [3] A. Raistrick, *op. cit.*, p. 25.

[4] *Reflections on various subjects*, p. 66. Late in the century Roe & Co. moved their copper smelting works from Liverpool to Neath because of the high price of labour and coal at Liverpool. A. H. John, *The Industrial Development of South Wales*, p. 28. It is not certain, however, that the lower wages of South Wales were the chief factor.

area. In Lancashire it was possible to distinguish three manufacturing regions: one to the east concerned with woollens, another to the west and south concentrating on linens, and a third lying between these two (with a south-eastern extension) occupied with fustians.[1] The West Riding presented a similar differentiation: woollens (or cloths) were made about Leeds and Wakefield, worsteds (or stuffs) a little to the west about Halifax and Bradford. In the metal region of the west midlands it was the same: parts of Shropshire and Worcestershire, close to the Severn, produced pig and bar iron; the elaboration of these was the chief business of the area to the east centred on Birmingham.

In each industrial province there was a growth in the number and size of the towns. Some of these were administrative, ecclesiastical, or trading centres which had drawn to themselves building workers, craftsmen, and servants to minister to the needs of residents whose incomes had little to do with industrial production. But others were created by manufacture. In earlier centuries strict limits had been set to the expansion of towns (other than seaports or river ports) by the amount of food that could be grown in the area from which, with existing means of road transport, it was possible to draw in supplies. But in the eighteenth century, turnpikes and canals widened this area. The urban concentrations that appeared after 1760 or 1770 were the result of improved means of communication, no less than of the growth of manufacture. The new towns, however, generally owed their inception to the advantages afforded by localisation of industry. The early corn-mills and fulling mills had often been established in rural areas, but the workers who gathered round them were sometimes the progenitors of a large urban group. The late Dr. W. B. Crump[2] has described how, in the eighteenth century, the fulling mills in the Yorkshire valleys drew from the hillside hamlets one textile process after another: first scribbling, next spinning, then finishing, and (much later) weaving. The fullers had often thrown a bridge across the stream and built cottages for their workers: the mill and the bridge were the nucleus of a village, which, as time went on, developed into a town. In other parts of the country whenever a new industry that required power entered a region, it tended to settle, not in the countryside but in

---

[1] A. P. Wadsworth and J. de L. Mann, *op. cit.*, p. 79.
[2] W. B. Crump and G. Ghorbal, *History of the Huddersfield Woollen Industry*, pp. 16-17.

an existing town. When, about 1717, Sir Thomas Lombe set up his mill for throwing silk he found his site in the town of Derby; and, when in the 'forties and 'fifties the silk industry spread, it took over and adapted corn mills or fulling mills in urban centres such as Manchester, Stockport, and Macclesfield. The engines of Thomas Newcomen were devised primarily for the draining of mines, but many of them were erected at places that were already, or were soon to become, towns. 'Perhaps there are not more than two or three throwing mills in the Kingdom', wrote an essayist[1] in 1752, 'nor above an hundred fire-engines, or a thousand water-engines (for these are rare except in towns).' When, in the last quarter of the century, water-power was applied to the spinning of cotton, some of the new factories were built in the country, in hamlets like Cromford, Mellor, and Styal. But others were set in or near urban communities, as at Chorley, Bury, and Holywell; and when steam-power was harnessed to cotton spinning, nearly all the new mills were in towns.

Many of the towns had their own particular process or product. In the metal area of the West Midlands, Birmingham specialised in guns, swords, buttons, buckles, and other 'toys' made of steel or brass. Wolverhampton and Willenhall made locks, Walsall bridles and bits, Dudley and Tipton nails, and so on. Sheffield concentrated on knives, razors, and files; scythes and sickles were made in the adjoining villages, and nails in the nearby parts of Derbyshire. Many textile towns produced fabrics that bore their names: some gave special attention to combing, others to weaving, and so on. Each industrial region had at least one large town that exercised quasi-metropolitan functions. Such regional centres drew to themselves the more skilled of the workers: they specialised in the lighter and more valuable products, and, like London, took a special interest in the finishing processes. Norwich, Leeds, Manchester, Sheffield, Birmingham, Bristol, and Exeter offer examples.

Once an industry had settled in a region there was every reason why it should stay there. Subsidiary trades grew up to minister to the principal occupation: the making of spindles, reeds, combs, and cards in the textile areas, of grindstones and crucibles at Sheffield, of shoemakers' knives and awls in leather-working centres such as Kendal, of stocking frames at Nottingham, and needles at Long Crendon. Manufacturers, who competed in the market, co-operated

[1] *Reflections on various subjects*, p. 32.

for common ends: 'public mills' for scribbling and fulling were common in the West Riding, and societies to encourage invention and improve technique arose in many parts of the country. Weekly markets and annual fairs dealt in the raw material and finished products of local industries. Manchester had its Exchange as early as 1729; Leeds set up a White Cloth Hall in 1711 and a Mixed Cloth Hall in the 'fifties; and most other centres of woollen and worsted manufacture followed suit. Reputations were established: the fact that a razor or knife was made at Sheffield, or a file at Warrington, gave it an advantage over one made elsewhere. The Cutlers' Company of Hallamshire exercised supervision of trade marks, and, after 1773, Birmingham had its Assay Office. Lines of communication were adapted to the needs of regional trade: specialised dealers, packers, and carriers appeared. Local banks adjusted their practices to the needs of local industry. There were often regional associations, like those of the northern coal-owners and the Midland and Yorkshire ironmasters, to regulate prices or wages. No producer lightly excluded himself from the external economies offered by localisation.

II

Industrial undertakings had varied forms. The skilled artisan who worked at home, or in a nearby shop, was still very common. When he owned his tools and stock-in-trade and sold his product on the market, he was clearly the owner of a business concern. Often he employed an apprentice and one or two journeymen, and sometimes he put out work to other men in the same trade. He flourished wherever close attention to detail and the maintenance of quality were of special account: in tailoring, cabinet-making, and other crafts of the metropolis, in the cutlery trades of Sheffield, and in sections of the woollen industry of the West Riding. Easy conditions of credit (in the first half of the century, in particular) favoured the small-scale producer. The master builder who put up the houses of all but the well-to-do was a working bricklayer or carpenter. He obtained a lease of a small plot of land, at a nominal rent for the first year or so, and, as building proceeded, replenished his resources by mortgaging the unfinished structure. He was thus able to pay for the services of masons, plumbers, slaters, glaziers, and plasterers, and maintain himself until the house, or row of houses, was ready for sale. In the same way, working shipwrights could get timber on credit from merchants, and funds for the payment of wages by giving a lien on the vessel in course of construction. Most shipyards, except those serving the Navy or the East India Company, were in the hands of men whose fortunes were slender.

In the so-called domestic industries there was little technological compulsion to large-scale undertakings. The appliances were small enough to find a place in the kitchen or bedroom of a cottage. In some branches of textiles raw materials and machines were relatively cheap; credit was easy to come by; a piece of narrow cloth could be woven in a week or two; and there were developed markets close at hand, where small quantities of fabric could be sold to merchants who saw to the finishing processes. In such circumstance the independent clothier-weaver might thrive. The homestead on the edge of the Yorkshire moors (where coal was accessible and springs of soft water abounded) was often a miniature industrial establishment. The 'house' or the 'great chamber' was cluttered with card stocks and spinning wheels, and the 'little chamber' was fitted with a loom. Attached to the building there was often a cloth-dresser's shop and a dye-house, and close at hand a tenterfield. Others besides members of the family might be employed. The small cluster of buildings sometimes included a tiny warehouse or 'takin'-in-place', to which men and women came each week to collect materials to spin and weave in their own cottages, the upper floors of which, with their long-mullioned windows, served both as bedrooms and workshops. But generally the work was done under the eye of the master, who divided his time between farming, weaving, supervising production, and attending the market at Halifax or Leeds.[1]

On the other hand, there were branches of the textile industries where conditions were unfavourable to the small producer. Sometimes materials had to be brought from remote places; some fabrics required a mixture of yarns; it might take a long period to produce a piece of broadcloth; the machine itself, as in the case of the Dutch loom, might be expensive; there might be no local market for sale of the cloth. When such was the case, the large merchant clothier prevailed. A small man might make and market a piece of kersey: he could hardly do the same for a piece of worsted or a variety of smallwares.

The greater part of the output of the textile industries came from the large undertakings. The territory covered by a single firm was

[1] W. B. Crump and G. Ghorbal, *op. cit.*, *passim.*

Manufactures

often extensive. The Norwich draper, the Leeds clothier, the Manchester warehouseman, and the Nottingham hosier, each, like a spider at the centre of a vast web, gave out material and drew in finished or semi-finished goods from hundreds, or even thousands, of spinners and weavers. One objection to the setting-up of the woollen industry near the coast was that the master manufacturer required 'a large circumference of country, for giving out the raw material and taking back the yarn or worsted: of which circumference the habitation of such masters is properly the centre'.[1] It would be wrong, however, to think of the labour force as distributed in neat geometrical patterns. If a manufacturer were accustomed to travel regularly to a port or market he might give out material all along the route. If he moved his headquarters he might continue to provide employment in his native town, as Samuel Oldknow did in Anderton after he had transferred his business to Stockport and Mellor.[2] But apart from such special cases, it took so many spinners to supply the needs of a weaver that it was often necessary to go far afield for yarn. According to Defoe 'the weavers of Norwich and of the parts adjacent, and the weavers of Spitalfields in London . . . employ almost the whole counties of Cambridge, Bedford, and Hertford; and besides that, as if all this part of England was not sufficient for them they send a very great quantity of wool one hundred and fifty miles by land carriage to the north, as far as Westmorland, to be spun; and the yarn is brought back in the same manner to London and to Norwich.'[3] It is difficult to believe that much wool can have been sent to a sheep-farming county. But there are other witnesses to the extent of the area over which a merchant manufacturer might cast his net. A producer of silks declared that, before 1762, he had employed people in London, Gloucestershire, Dorset, and Cheshire.[4] In the 1770's the clothiers of Halifax were putting out work, across the Pennines, in Lancashire.[5] About the same time other Yorkshiremen employed jersey spinners at Wilmslow

---

[1] *A Letter to the Author of a Brief Essay* . . . *on the Advantages and Disadvantages which attend France and Britain with regard to Trade* (1751), p. 13.

[2] G. Unwin and others, *Samuel Oldknow and the Arkwrights*, pp. 49-54.

[3] From *The Complete English Tradesman*. The passage is reproduced in A. E. Bland, P. A. Brown, and R. H. Tawney, *English Economic History Select Documents*, pp. 484-5.

[4] M. D. George, *op. cit.*, p. 185.

[5] J. Bischoff, *Comprehensive History of the Woollen and Worsted Manufactures* (1842), p. 185.

in Cheshire.¹ And towards the end of the century it was observed that in the county of Huntingdon women and children were spinning yarn for the Norwich and north-country (presumably Yorkshire) markets.

The numbers employed by a concern in the textile industries must often have been far greater than those brought together at a large colliery or ironworks. The two brothers at Blackburn who claimed, in 1736, to employ 3,000 people, the Warrington sailmaker who declared that, before 1750, he had given employment to 5,000, and the silk manufacturer, referred to above, who put the number of his employees at 1,500, were probably not guilty of wild exaggeration.² Few later factory masters could have sustained such a claim. It is less easy to estimate the capital involved. The dwellings of the workers in which spinning and weaving were done must be considered, in part, as industrial plant. To what extent the capital they represented was provided by landowners, master manufacturers, or the occupiers themselves, is a matter to which local historians might well direct their enquiries. Each industrial household had its tools and machines. Sometimes the textile worker, like the miner or carpenter, provided his own appliance: sometimes he rented it from the employer or middleman. But whatever the provenance, the investment in fixed capital by those belonging, in some sense, to a concern, must often have been substantial. Further capital was embodied in stocks of material and goods in process of manufacture. Production might extend over many months: there was often a long interval between the completion of the goods and their sale, and sometimes a still longer one between sale and receipt of payment; instances are to hand of clothiers who had to wait as long as two years for their money. It was in circulating, rather than fixed, capital that the textile manufacturer held the greater part of his assets, and in some cases these must have run into scores of thousand pounds.

Similar variations of scale and organisation are to be found in the metal-working industries. Some artisans bought rod iron from one merchant and sold the finished wares to another. Even when they returned their product to the man from whom they had obtained the material, they thought of themselves not as servants or outworkers, but as independent producers. In February, 1794, William

---

[1] Samuel Finney, *An Historical Survey of the Parish of Wilmslow.*
[2] A. P. Wadsworth and J. de L. Mann, *op. cit.*, pp. 211, 404; and M. D. George, *op. cit.*, p. 185.

Harrison of Chester wrote Peter Stubs of Warrington for information as to 'the workmen's prices of nails and what they give their masters for iron'. The wording is significant of the status of the nailmakers in this area. Many of the smiths and file-cutters who worked for Stubs also took orders from others: they received a price for their product and not a wage for their labour. But in the midlands the nailmakers and other workers in iron are nearly always referred to not as craftsmen or artisans, but as workers: they had little sheds attached to their cottages, but they were wage-earners paid by the piece. The men who employed them were wealthy ironmongers or manufacturers with warehouses at Birmingham or Dudley; a single one of these might put out work to several hundred cottagers. The frequent protests of the workers that they had to accept payment of wages in truck, and of the employers that the workers were given to rebellions and tumults, are eloquent of the gulf that divided master from man.[1]

It was obviously impossible for the merchant clothier, hosier, or ironmonger to have direct contact with all who worked for him. Intermediaries were necessary. Mr. Wadsworth has shown how, in the late seventeenth century, a fustian dealer, Thomas Marsden of Bolton, employed a country manufacturer, Hugh Pickering, to put out cotton and yarn in the area about Darwen. Pickering was debited with the materials and credited with the wages he had paid to the spinners and weavers. He received a commission on each piece of fustian or pound of weft delivered to Marsden. But, though he was thus an agent, he was also a fustian-maker on his own account. It is probable that in the eighteenth century the same combination of activities persisted, as, indeed, it did in the case of the bag-men in the hosiery trade of the nineteenth century. But most large merchant manufacturers employed persons to travel from place to place to give out materials, collect finished work, and pay wages, not at the cottage door but at depots set up in various parts of their industrial provinces. Often the putter-out was a man in an entirely different occupation. In 1794 Samuel Gaskell & Co. asked Peter Stubs to send two dozen whipsaw files to Samuel Barrett of Woore, adding that 'he puts out spings for us, and when there is any money due to him (we) will pay you for them'. Evidently Barrett was a craftsman, perhaps a carpenter: his function as a putter-out

---

[1] References to such grievances can be found in the pages of the House of Commons' Journals at many separate points of time.

was clearly subsidiary to some other calling. In rural areas, as in Oxfordshire, the barn of a farmer or the outbuildings of an innkeeper sometimes served as a local warehouse; and in Huntingdon, it was said, yarn was put out 'by the generality of the country shopkeepers'. It is easy to see why the putting-out system was often associated with truck.

Generally the worker had to do his own fetching and carrying, to and from either a local warehouse or the headquarters of the clothier or merchant. Now and then use was made of the farmer's cart or the carrier's wagon. But on the roads of the north large numbers of weavers were to be seen bearing yarn in packs on their backs, or heavy rolls of cloth under their arms. The distances covered were often as great as most men would care to traverse in a day. The weavers of Farnworth had to tramp eight miles to Manchester and back again: those of Grindleton made each week, a ten-mile journey to Barnoldswick.[1] In other domestic industries it was the same. No evidence has been found of the existence of local depots in the metal-working industries—though this must not be taken as a statement that none existed. The smiths and nailors of South Staffordshire had to travel long distances between their homes and the warehouse of the ironmonger in Birmingham or Dudley, carrying heavy iron rods or bags filled with nails and other wares.[2] A filemaker, Carolus Charles of Aughton, covered twelve miles on foot to Green's warehouse at Liverpool; and many of the scattered toolmakers employed by Peter Stubs were obliged to travel similar distances to the White Bear at Warrington. As in most under-developed countries today, a large part of the energies of poor men and women was given to transport. It is said that in the hosiery trade of the east Midlands as much as two and a half days a week might be taken up in getting orders and material, returning finished work, and collecting wages. When account is taken of the time spent in amusement in the town, or solace at the wayside inn, the estimate may not seem excessive.

Enough has been said to indicate that in many industries the boundaries of the firm were ill-defined, the structure loosely organised, and the workers remote from the centre of control. In the textile trades, in particular, there must have been thousands of workers who never set eyes on their employer. The

[1] A. P. Wadsworth and J. de L. Mann, *op. cit.*, p. 394n.
[2] Samuel Timmins, *Birmingham and the Midland Hardware District*, p. 86.

Manufactures

notion that the coming of factories meant a 'depersonalisation' of relations in industry is the reverse of the truth.

III

The specialisation of areas described above was matched by a growing specialisation of work. An economic historian once asked why Adam Smith had drawn his illustration of the effects of the division of labour from his 'silly pin manufactory' when a few miles away from his home stood the great Carron ironworks. The answer is obvious: Adam Smith was anxious to isolate the results of the application of his celebrated principle from those of the use of machinery and power. The pin trade employed only simple appliances: it was almost ideal for his purpose. Most English industrial establishments came somewhere between the extremes represented by the pin works and Carron. The economies they were able to achieve were based partly on the specialisation of labour and partly on that of capital. The two were, indeed, interdependent: some degree of division of labour was essential to the introduction of machinery; some application of capital in fixed forms was a condition of any high degree of division of labour. Dr. George has shown how in the watch trade of London 'cutting engines', devised in the late seventeenth century, simplified the work of the craftsmen and made it possible to divide and sub-divide processes. By 1747 division of labour was intense: there were wheel-forgers, wheel-cutters, pinion makers, spring makers, cap and stud makers, jewellers, engravers, chasers, and enamellers, beside the finishers who assembled the varied parts and the key makers and chain makers who provided accessories.[1] It was the same with the heavy industries. At the beginning of the century most of the miners who worked coal by the longwall method were either getters or drawers. By the end of it a growth in the size of the pits, and the use of capital appliances, had created specialised occupations including those of holers, hammermen, remblers, punchers, loaders, trammers, hangers-on, and banksmen. In the areas where working was by bord and pillar the simple classification of hewers and putters had similarly given way to a varied nomenclature: there were trappers, trammers, half-marrows, marrows, headsmen, put-and-hewers, onsetters, brakesmen, and banksmen.[2] In the iron industry alongside the furnace-keeper, filler,

[1] M. D. George, *op. cit.*, pp. 173-5.
[2] T. S. Ashton and J. Sykes, *op. cit.*, ch. II.

forgeman, and slitter, there appeared the puddler, roller, and moulder. The early brass founder was a man who performed varied operations: by 1770 pattern-making, mould-making, casting, and finishing were becoming distinct occupations, and new processes of stamping and pressing were bringing into being still other classes of specialised labour.[1]

The woollen industry had long been marked by a high degree of division of labour. Before the eighteenth century a process known as scribbling had been introduced as a prelude to carding, and, after the invention of the jenny, another known as slubbing was interposed between carding and spinning. The innovations of Arkwright and Crompton brought a minute subdivision into cotton spinning: wages-books of the 1790's contain names of occupations which only those highly versed in technology are able to interpret.

No attempt will be made to describe the course of invention that led to a parallel specialisation of capital. The outlines of the story are well known, though most of the accounts lay undue stress on the achievements of individuals like Darby, Watt, Cort, Kay, Hargreaves, Arkwright, and Crompton, and too little on those who, by adding one tiny device to another, or modifying this or that process, prepared the way for such men. There can be little doubt that, in the eighteenth century, the impulse to contrive was widespread: a writer of 1752 spoke of 'the infinite numbers daily inventing machines for shortening business';[2] and about the same time another declared that 'at Birmingham, Wolverhampton, Sheffield and other manufacturing places, almost every manufacturer hath a new invention of his own, and is daily improving on those of others'.[3] Various explanations may be given of this outburst of ingenuity. Englishmen had settled their political and theological differences and were turning their energies to practical ends. The scholarly curiosity of the sixteenth and seventeenth centuries had filtered down to humble people whose forebears could have had little time for speculation or experiment. Schools and academies were spreading enlightenment: it is perhaps not without significance that

---

[1] H. Hamilton, *The English Copper and Brass Industries to 1800*, pp. 266-7; G. C. Allen, *The Industrial Development of Birmingham and the Black Country*, pp. 17-18.

[2] *Reflections on various subjects relating to Arts and Commerce* (1752), p. 31.

[3] 'Sir John Nickoll', *Remarks on the Advantages and Disadvantages of France and Great Britain with respect to Commerce* (1754), p. 21.

Arkwright should have made his first spinning frame in the grammar school at Preston—though the influence cannot have been comparable to that of the University of Glasgow on James Watt. Newspapers and books spread knowledge and aroused aspirations. The trend in literature was towards reality, clarity, and common sense. If *Pilgrim's Progress* was the epic of the ordinary man of the seventeenth century, *The Life and Strange Surprising Adventures of Robinson Crusoe* was that of his successor of the eighteenth. This may have led many boys to far-off seas, but it must also have stimulated improvisation by some whose adventures were confined to the home or the workshop. The decline of strict apprenticeship and corporate regulation of industry released native talents. More ample supplies of capital, reflected in low rates of interest, reduced the cost of experiment and the construction of new appliances. And (though this is far from certain) the protection afforded by the law of patents may have encouraged contrivance. None of these influences, however, provides a fully satisfying explanation. Why did one star shine more clearly in the thirteenth century and another in the sixteenth? Why Italian art, or Elizabethan drama? The historian cannot give a sure answer.

The process of invention was rarely simple. It involved the realisation of a need, the conception of an idea, the devising of a process or the making of a model, and then the long, disheartening business of adapting the innovation to the conditions of the workshop or the factory. Most industrial discoveries were the product of joint enterprise. Without the scholarly learning of Black and the technical skill of Wilkinson, James Watt could hardly have created the steam engine; if Arkwright had not had a predecessor, Lewis Paul, and contemporaries like Thomas Highs and Jedediah Strutt, it is doubtful whether his name would be remembered today. Few men of mechanical genius had the resources or acumen to carry their projects to success, and there are more than a few stories of frustration. But the idea that inventors were singled out for misfortune is a popular error. Many of them found partners with the capital, practical sense, and enterprise they themselves lacked. Too often have the men who supported an innovation been regarded as exploiters of other men's brains: too rarely has account been taken of those who lost all they had in the venture. Entrepreneurs of the type of Roebuck, Garbett, Boulton, and Crawshay were no less essential than originators, like Watt and Cort, to the process that transformed the technique of industry.

This is not to say that things might not have been better ordered for the inventor. Under the Statute of Monopolies of 1624 the Crown was empowered to make grants of privilege, for periods not exceeding fourteen years, 'of the sole working or making of any manner of new manufactures within this realm to the first inventor or inventors of such manufactures'. It was held that without such protection there would be little incentive to discovery, and that men would seek to conceal their devices: the registration of a patent gave at least a little information to the public and ensured that all knowledge of the process should not die with the author. Universal though it has now become, this method of recognising ingenuity had (and still has) serious defects. It was not easy to identify the first inventor. For many decades it was possible for a man to obtain a patent for introducing to the country a device that was not of his own creating, or for a mere idea or project. And, although early in the eighteenth century it was decreed that applicants should 'particularly describe' the nature of their inventions, many succeeded in passing specifications that were deliberately vague, and so were enabled to ring off for themselves large fields of potential profit. An arrangement devised to stimulate what we now call research had the effect of refusing to many the opportunity of exercising their talents. Moreover, the procedure of obtaining a patent was involved. The cost was high and there was a risk of suits in the courts of law. A man of small means often thought it best not to risk burning his fingers.[1]

It was argued, with force, that the interests of the nation would have been better served by a public award of premiums and honours. But though in a number of instances, including those of Thomas Lombe, John Harrison, and (somewhat belatedly) Samuel Crompton, the government made substantial grants, there was a fear that extension of the practice to less clear cases of merit might lead to corruption. For this reason some advocated a system of rewards by private agencies. The Society for the Encouragement of Arts, Manufactures, and Commerce was established in 1754 largely for this purpose; and other bodies, like the Manchester Committee for the Protection and Encouragement of Trade, of 1774, had among their objects opposition to patents and the substitution of other means of recognising ingenuity.[2] The awards and medals distributed by the Society of

---

[1] Witt Bowden, *Industrial Society in England towards the end of the Eighteenth Century*, ch. I.
[2] A. Redford, *Manchester Merchants and Foreign Trade*, 1794-1858, ch. I.

Arts were, however, small bait, and the fact that the model or appliance submitted became the property of the Society must have discouraged applications for grants. Though the Manchester spinners were wholehearted in their opposition to the monopoly of Arkwright, they showed less zeal in collecting subscriptions for other inventors. And so the patent system continued. The question whether, on balance, it stimulated or impeded the progress of technique admits of no clear answer. James Watt deserved well of his fellows; but when, in 1775, Parliament extended the patent, granted him in 1769, for a further twenty-five years, it gave great power to a man whose ideas were, before long, to become rigid. Watt refused applications for licences to make engines under his patent: he discouraged experiments by Murdoch with locomotive models; he was hostile to the use of steam at high pressure; and the authority he wielded was such as to clog engineering enterprise for more than a generation. If his monopoly had been allowed to expire in 1783 England might have had railways earlier.[1] If a similar privilege had been extended to Arkwright—if, indeed, his wide patents had not been annulled in 1781-5—it is at least possible that a dead hand might have rested on the cotton industry also, and that forces tending to raise the standard of life of the poor would have been stifled. But speculations as to what might have been are vanities.

If (as is very doubtful) the number of patents may be used as an index, inventive fertility increased rapidly. And, as with other series, the growth was specially marked in the last two decades of the century. In the ten years, 1771-80, the total number of patents granted was 297; in 1781-90 it rose to 512, and in 1791-1800 to 655. According to Schumpeter, the process of invention is most active in periods of depression: in his view it is the pressure of wants that gives rise to ingenuity. It is impossible to prove that he was wrong. But it is worth while to notice that the number of new patents rose in each period of optimism and mounting output, and declined when the reverse conditions prevailed. In 1782, when a disastrous war was drawing to its end, only thirty-nine patents were taken out: in the following year of peace the number rose sharply to sixty-four. The year of depression, 1788, produced forty-two patents, the year of boom, 1792, no fewer than eighty-five. With the outbreak of the French war the figure fell to forty-three, and though in 1796 it

[1] Though, in view of the war and high rates of interest, probably not very much earlier.

reached seventy-five, the crisis of the next year brought it down to fifty-four. In 1800 the peak of the century was registered with ninety-six new patents. It may, of course, be that men brooded over their devices in the bad years and waited for better times before announcing their discoveries. But it is at least equally possible that ideas themselves germinated more readily in a benign and invigorating climate. It should be added, however, that considerations of cost probably influenced applications: in nearly every year when the number of new patents rose sharply rates of interest were low, and in every year of marked decline, these were high. It seems likely that the fluctuations were determined by factors similar to those that bore on the number of acts of enclosure, turnpike trusts, canal companies, and other projects.

IV

From the point of view of the technologist the innovations were directed to varied ends: making use of the force latent in water at a given elevation; harnessing atmospheric pressure and the expansive power of steam; releasing, in the form of heat and energy, the properties of coal; giving new applications to the wheel and the lever; reducing friction; and providing conditions for chemical reactions to produce the acids and alkalis necessary to industry. It would be easy to extend the list. From the point of view of the economist, they represented a replacement of scarce by less scarce resources. In many instances it was a matter of substituting relatively cheap capital in the form of appliances for relatively dear labour; and these have been given undue prominence in the text books. It has even been asserted that a desire to reduce wages was the principal motive for the introduction of machinery. In fact, the workers set free by the devices introduced at this period were generally poorly paid labourers rather than mechanics, spinners rather than artisans. The early machines were of too elementary a nature to be substitutes for high skills: that alone may explain what, at first sight, appears an anomaly. But there is a further consideration: low wages are not synonymous with low labour costs. In coal mining the chief cost was not that of the relatively well-paid man at the coal face, but of the lower-paid loader, putter, winder, and waggoner. It is not surprising therefore that ingenuity and capital should have gone to the improvement of the underground ways: to laying rails of wood or iron, providing conditions in which ponies could take the place of boys and women as beasts of burden, and atmospheric or steam engines that of the men at the pumps. In the early years of the century there were in the mines of Northumberland and Durham quite as many putters or barrowmen as hewers: by the beginning of the following century the number of men at the face (the best paid of the workers) far exceeded those who transported coal underground.[1] The same thing occurred in the textile industry. Here the chief cost was that of the poorly paid women spinners, who greatly outnumbered the fullers and weavers, and the employment of whom, in their scattered homes, involved high expenditure on transport and book-keeping. It was not high wages but high labour costs that led to the introduction of the devices of Hargreaves, Arkwright, and Crompton. It is true that, long before these came into being, James Kay had turned his attention to improvements in weaving. But one object, and result, of the flying shuttle was to make it possible for the craftsman to weave a broad piece of cloth without an assistant. The progress of the device was far less rapid than that of the jenny, the frame, and the mule; but it had a similar effect in eliminating the less well paid of the weavers.

Frequently, however, inventors aimed at economy of natural resources. One of the chief anxieties of the industrialists of the early part of the century related to the supply of water. If that failed, the mills that ground corn and malt, fulled cloth, slit iron, or ground rags to make paper, were brought to a halt. Very large sums were invested in dams, ponds, culverts, troughs, and water-wheels, and the annual cost of maintenance was high. One of the most important developments of the early part of the century was the substitution of the over-shot for the under-shot wheel.[2] By its use a small volume of water could supply as much energy as a far larger volume with the other device. When the atmospheric engine was employed (as it was at Coalbrookdale) to throw back water from below the mill race, so that it could be used again and again, the saving was spectacular. And when the rotary engine of James Watt replaced the water-wheel as a prime mover the gain was incalculable. Similar economies were made by the engineers who substituted locks for dams and weirs on the rivers: previously large volumes of water had gone to waste whenever a 'flash' was provided for vessels moving

[1] T. S. Ashton and J. Sykes, *op. cit.*, p. 68.
[2] It is impossible to assign a date. But about 1727 it was said: 'overshot watermills are become most general (especially in the Northern parts of England)'. See J. E. Thorold Rogers, *op. cit.*, vol. VII, pt. II, p. 621.

up or down stream. Brindley saved water when he lined the bed of his canals and aqueducts with puddled clay; and a much larger saving was realised by the use of the pound-lock with double gates. All these, in varying degrees, are instances of the substitution of capital for natural resources, though some of them were economical also of other factors of production.

In other cases the effect of an innovation was to liberate capital. The steam engine not only economised water but made it possible to dispense with costly hydraulic works: an industrialist who took a site on the bank of a canal could get all the water he needed for his boilers and condensers at little or no expense.[1] When iron took the place of copper and brass in engine parts, of stone or brick in bridges,[2] and of timber for the beams and pillars in factories,[3] there was a considerable capital saving. Tools and machines made of wrought iron were lighter, more efficient, and more durable, than those made of wood. It was this kind of economy Adam Smith had in mind when he wrote that 'all such improvements in mechanics, as enable the same number of workmen to perform an equal quantity of work, with cheaper and simpler machinery than had been usual before, are always regarded as advantageous to every society'.[4]

Much of the innovation was directed to the economy of horse-power (in the literal sense of the term). The Newcomen engine applied to the working of pumps, the wheeled corf in the underground ways, the rotative steam-engine that wound the coal up the shaft, wagon-ways on the surface, such as those of the Duke of Norfolk at Sheffield—all these led to substantial reductions in the use of horses in or about mines. The saving of horses by the new forms of inland transport is too obvious to require further illustration. But the same thing occurred in manufacturing industry. The primitive factories of Lewis Paul, and, much later, of Edmund Cartwright, derived their power from horses, and the same was true of ropewalks, glass works, and other establishments for which water-power was either unavailable or unsuited. Edmund Halsey (the uncle of Ralph Thrale) bought blind horses to work the pumps and mills of his brewery at Southwark. But towards the end of the century at nearly all the great London breweries a steam pump was used to

---

[1] The Acts authorising canals in the 'nineties sometimes required the companies to supply water for condensing free of charge.
[2] The first iron bridge was erected over the Severn in 1779.
[3] A development of the last years of the century.
[4] *The Wealth of the Nations*, vol. I, p. 252.

raise water, and an Archimedean screw to carry malt to the top of the building, whence the liquor flowed down through the successive stages of mashing, boiling, fermenting, and cooling to vats in the basement.[1] Innovations concerned with the substitution of other prime movers for horses might incidentally save labour, but the essential economy was in capital.

The capital released might go into the construction of a larger number of machines, but it might, equally well, take the form of goods in process of production. In continuation of his theme, Adam Smith says:

'A certain quantity of materials, and the labour of a certain number of workmen, which had before been employed in supporting a more complex and expensive machinery, can afterwards be applied to augment the quantity of work which that or any other machinery is useful only for performing. The undertaker of some great manufactory who employs a thousand a-year in the maintenance of his machinery, if he can reduce this expense to five hundred will naturally employ the other five hundred in purchasing an additional quantity of materials to be wrought up by an additional number of workmen. The quantity of that work, therefore, which his machinery was useful only for performing, will naturally be augmented, and with it all the advantage and conveniency which the society can derive from that work.'

On the other hand, innovation sometimes resulted in a transformation of circulating into fixed capital. In the early part of the century a high proportion of the wealth of the nation was embodied in stocks of material in the hands of merchants, dealers, and domestic workers. The longer the time occupied in production the greater the capital cost. It has been remarked elsewhere that in no respect did the men of the eighteenth century differ more from their predecessors than in their attitude towards time.[2] In his advice to the young, Sir John Barnard (who had grown to affluence as a merchant and had served as Lord Mayor of London) wrote, 'Above all things learn to put a due value on *Time*, and husband every moment as if it were to be your last: in Time is comprehended all we possess, enjoy, or wish for; and in losing that we lose them all.'[3] It would be foolish to treat

---

[1] Peter Mathias, 'Industrial Revolution in Brewing', *Explorations in Entrepreneurial History*, vol. V, pp. 213-14.
[2] T. S. Ashton, *The Industrial Revolution*, p. 99.
[3] *A Present for an Apprentice*. By a late Lord Mayor of London (1741), p. 20.

the sententious utterances of successful men as evidence of the attitudes of ordinary people; but no one can read through the correspondence of the manufacturers of the period without noticing the stress laid on punctuality and the exasperation caused by any unnecessary delay. If the time taken up in production and delivery could be shortened resources would be available for other purposes. Hence the importance of the innovations in bleaching: under the older methods cloth had had to be exposed to the action of weak acids and sunlight for several months; with the introduction of chlorine in the 'eighties (and of bleaching powder in the 'nineties) the process could be completed within the course of a few days. Every contrivance that made wheels turn more quickly, every increase in the speed of transport, was economical of materials. A good deal has been written on the sources of the capital that was embodied in the early factories: on the extent to which this came from personal savings, undistributed profits, bank advances, and so on. In an ultimate sense much of it was not a new creation at all. Some of the early factory masters simply took over and adapted existing corn mills, fulling mills, or warehouses. And, though most of them had an immediate problem of finance, the resources required to extend, equip, and maintain their premises were already within the industrial system. The speeding up of production and distribution by the new machines and new means of transport made it possible to transmute circulating into fixed capital. The process is at the centre of what is called the industrial revolution.

It is not suggested that each innovation can be labelled and classified according to the particular kind of saving it produced. Some of the inventions (the steam-engine is the best example) were economical of natural resources, capital, and labour concurrently. The purpose of these remarks is simply to point out that the industrial discoveries involved new combinations of productive factors, and that it is by no means true that their sole object was to economise labour. It should be unnecessary to add that an innovation might increase, rather than decrease, the proportion of labour required in production. But a period when natural resources were abundant and capital generally cheap is unlikely to offer many instances. The efforts of Poor Law authorities to find employment for women and children in spinning on old-fashioned machines ran counter to the temper of forward-looking industrialists. But at the end of the century fears of over-population led to various proposals which, if put

Manufactures

into effect, would have replaced capital by labour. Such, indeed, was the result of the innovation made, a little later, by Thomas Telford. His method of dressing the roads with small angular fragments of stone, so as to make the surface impervious to water, was prodigal of labour. Crushing machinery had long been used by the ore miners of Cornwall and other areas, though whether or not it could have produced road metal of the kind required it is impossible to say. Certain it is that, in the years of depression that followed the wars with the French, thousands of men, women, and children were employed, near Bristol in particular, in breaking up rock with hammers, to provide material for use by the turnpike trusts. Public relief works, at least, afford examples of innovations that were emphatically not labour-saving in their effects.

V

In the early years of the century the largest units of production were those controlled by public authorities. Writing in the early 1720's, Defoe describes the elaborate organisation of the royal arsenal at Chatham:[1]

'The buildings here are indeed like the ships themselves, surprisingly large, and in their several kinds beautiful: The ware-houses, or rather streets of ware-houses, and store-houses for laying up the naval treasure are the largest in dimension, and the most in number, that are any where to be seen in the world: The rope-walk for making cables, and the forges for anchors and other iron-work, bear a proportion to the rest; as also the wet-dock for keeping masts, and yards of the greatest size, where they lye sunk in the water to preserve them, the boat-yard, the anchor yard; all like the whole, monstrously great and extensive, and are not easily describ'd. . . .

'The building-yards, docks, timber-yard, deal-yard, mast-yard, gun-yard, rope-walks; and all the other yards and places, set apart for the works belonging to the navy, are like a well-ordered city; and tho' you see the whole place as it were in the utmost hurry, yet you see no confusion, every man knows his own business; the master builders appoint the working, or converting, as they call it, of every piece of timber; and give to the other head workmen, or foremen their moulds for the squaring or cutting out of every piece, and placing it in its proper byrth (so they call it) in the ship that is

[1] D. Defoe, *op. cit.*, vol. I, pp. 106-8.

in building, and every hand is busy in pursuing those directions, and so in all the other works.'

Ambrose Crowley had his headquarters at Greenwich: it is possible that the great industrial organisation he built up owed something to this example set by public enterprise.

In a much smaller way the policy of local authorities led to aggregations of labour. The desire of justices of the peace to reduce the burden of the rates resulted, as early as 1705, in the setting up of workhouses where the poor and their children could be employed in spinning and other simple tasks; and by 1725 about 124 of these were in being.[1] They were owned by the parishes, or rather by unions of parishes, but were generally let to contractors, who undertook to maintain the paupers at so much a head. If they were rarely commercially successful that was largely because of the nature of the labour they employed. Some were little more than barns; but others, built in the form of a square, with workshops, stores, and offices, served as models for the privately owned factories that came later.[2] Side by side with them there existed, in many parts of the country, spinning schools, established by private benevolence, for the instruction of poor children. There can have been few regions to which the idea of the concentration of workers under a single roof was wholly unknown.

It was not only by example that the state encouraged the rise of large-scale productive units: the demands of the armed forces for guns, uniforms, boots, and other equipment told in the same direction. It was more convenient to place contracts with firms that disposed of sufficient labour, under supervision, to be able to give guarantees of quality and punctual delivery. Sail-cloth could hardly be made in cottages. But it was probably the size of the orders from the navy and the East India Company that led to its production in London in long buildings 'on the same system as ropeworks' in which large numbers of spinners and weavers worked side by side.[3] In the iron industry, firms like the Wilkinsons, the Walkers, and the Carron partners owed the size of their undertakings partly to government contracts for cannon; and in the woollen industry the supplying

---

[1] W. Cunningham, *The Growth of Industry and Commerce in Modern Times*, p. 574n.
[2] For a description of the workhouse at Tiverton, see M. Dunsford, *Historical Memoirs of Tiverton* (1790), pp. 358-9.
[3] M. D. George, *op. cit.*, p. 165. For another example of large-scale production of sail-cloth see A. P. Wadsworth and J. de L. Mann, *op. cit.*, p. 404n.

of cloth for uniforms and army blankets aided the growth of concentration by businesses like that of Benjamin Gott. It was so also in the provisioning of the forces: in 1781 Smeaton set up a great corn-mill, worked by water raised by Newcomen engines, to supply flour to the navy.

Fiscal policy, also, tended to produce concentration. Throughout the century the government relied largely on excise duties, the raising of which involved oversight of manufacture by officials. The small producer was injured more than his larger competitor by the cost of the licence; and he might have to suspend operations till the exciseman saw fit to visit his works, whereas in a large establishment an officer was on the spot at all times. The reduction of the number, and the growth of size, of undertakings producing malt, beer, spirits, glass, paper, leather, and printed fabrics must have owed something to this influence.

In industries not subject to encouragements and pressures from the state, the drawing of labour into places of common employment was the result of the localisation, division of labour, and technical innovations discussed above. As agriculture came to require more specialised workers, less spinning and weaving was done in farmhouses and cottages; and, indeed, landlords sometimes prohibited the practice. The growth of industry in towns was favourable to larger units. When a new manufacture was introduced to an area labour had to be trained and supervised; and if new appliances were required it was usual for these to be provided by the employer. To give a single example: when, about 1752, the weaving of Norwich stuffs was brought to Devonshire, a Tiverton clothier, Martin Dunsford, broke down the party walls of a row of houses opposite his dye-house to form a workshop, where seventy or eighty people were employed for weekly wages.[1] But shops of this kind had long been a feature of the production of cloths of quality in the West Country, for the Spanish and other high-grade wool was too expensive to be entrusted to scattered cottagers. Wherever there was serious danger of embezzlement, wherever goods might suffer in transit, wherever time was important, there was a tendency to draw spinners and weavers into places where supervision could be exercised.[2] In the silk industry of London there had been throwsters'

[1] M. Dunsford, *op. cit.*, p. 236.
[2] In the West Riding, Law Atkinson, who made use of Spanish wool, declared that it was in order to check embezzlement that he had taken to factory production. Crump and Ghorbal, *op. cit.*, p. 92.

shops in the seventeenth century, and, from its first days, smallware manufacture was carried on in establishments other than the home. For technological reasons the same thing was true of bleaching and the printing of fustians and calicoes: in 1762 a print works near Manchester was giving employment to between ninety and a hundred 'artists and working people'.[1]

The chief reason for concentration, however, was the application of power. Horses and water wheels were employed, before 1714, not only in fulling, but also in the earlier textile processes. There were linen-thread works at Manchester, powered by horses, in the 'twenties.[2] Reference has already been made to the way in which, in the West Riding, scribbling, slubbing, spinning, and even weaving, were gradually drawn to the fulling mills on the rivers. The most striking early instances of concentration based on power are, however, to be found in the silk industry. Between 1717 and 1721 Thomas Lombe set up at Derby a true factory 'half a quarter of a mile in length' with 26,000 wheels, and more than 300 workpeople; and in the three following decades, Italians and others who had been trained at Derby put up similar structures at Stockport, Macclesfield, Leek, and Congleton.[3] In the 'forties the spinning devices of Lewis Paul were housed in factories, worked by horses or waterpower, at Birmingham, Northampton, and Leominster. These were smaller in scale than the silk mills, but they must have done something to accustom people to the idea of concentrated production of yarn. The use of power was the dominant influence leading to integration: as Mr. Wadsworth has shown, there were growing up, in the 'seventies, many small factories in Lancashire in which carding was carried on in conjunction with spinning; and attached to these were sheds in which weavers, who still had to rely on their own arms for power, worked under supervision of the employer.

Hargreaves's jenny was a hand machine. But since it produced far more yarn than the wheel, it almost certainly led to a decline in the number of cottages in which spinning was done. Still more was this true of Crompton's mule, which was difficult to work and called for the labour of skilled men. There was a tendency for both devices to be concentrated in relatively large workshops. It was, however, the water-frame and carding machine that, after 1771, gave rise to truly large-scale units, first in the cotton, and later in the woollen

[1] A. P Wadsworth and J. de L. Mann, *op. cit.*, pp. 105-7, 131, 135-7, 308.
[2] *Ibid.*, pp. 107, 303.
[3] For further details see William Hutton, *History of Derby* (1791), p. 198.

industry. The factories set up by Arkwright, or under his licence, at Cromford, Belper, Milford, Birkacre, Holywell, and many other places, housed hundreds of machines, all deriving their power from a great wheel. Above all, they were integrated concerns in which the raw cotton passed through a series of operations till it emerged as finished yarn. It was the aggregation of processes, quite as much as the size of the factory, that aroused opposition—in Yorkshire from men who had long been familiar with, and had made use of, the so-called public fulling mills. And it was partly the dislike of adults for these buildings (which, if we may judge by those that have survived, were by no means the dark, satanic mills of popular legend) that led to their being staffed by children and young people.

Arkwright began at Cromford in 1771 with 300 employees: ten years later his labour force had trebled. At this time, there were about twenty water-driven factories: by 1790 the number had risen to 150. The application of Watt's rotative engine to carding and spinning brought about further increases in number and scale. Most of the steam mills were in the towns, and no limit was set to their growth by difficulties of obtaining and housing the workers. A steam engine could supply power to three or four separate establishments, but generally its effect was towards concentration. In the 'nineties the whole setting of industrial life in the north was transformed.

No less important, though less striking, changes occurred in the structure of other industries. In the production of iron, limitations of both fuel and power had for centuries led to a separation of refining from smelting. It is true that many ironmasters had controlled operations in both branches of the industry, but the furnaces and forges had rarely been on the same site. The application of steam power, and the introduction of puddling and rolling by Cort in 1783, removed the barriers to integration. In Staffordshire, Yorkshire, and —above all—South Wales, all processes, from the mining of coal and ore to the slitting of rods and the production of finished wares, were henceforth carried on in the same locality, by the same firm, and most of them in single establishments. Steam power was applied to mining, paper-making, brewing, distilling, corn-milling, and a variety of other trades. Iron, produced by Cort's process, took the place of wood in shafts, gears, wheels, and machines, and this, also, powerfully aided the growth of scale of the factory or works.

Often a craftsman or manufacturer took his sons or other relatives as partners. The initial capital came from the personal savings of two or three people, and year by year this was augmented out of profits. As the business grew in size and complexity it was sometimes necessary to allow an experienced employee to share in the control, or to invite a merchant, lawyer, or banker to become a partner, active or sleeping. And frequently, when the partnership was still small, it appended to the name of the head of the concern the imposing inscription '& Co.'—'imposing' (as an American reviewer said of a book) in both senses of the word. For, from first to last, most manufacturing establishments were essentially private, or family, concerns. Nowadays, the family business is often open to the charge of inertia. But in the eighteenth century it mobilised resources and developed enterprise as no other form of organisation could have done. The story of the Darbys, Wilkinsons, and Walkers in the iron industry and of the Arkwrights, Strutts, and Peels in the cotton industry, is sufficient to establish the point.

Sometimes partnerships were formed by a few men, not otherwise related, in order to obtain supplies for their separate concerns. In the second decade of the century Abraham Darby joined with three others to set up the Vale Royal furnace in Cheshire. Each contributed materials, skill, or money; and every Monday morning the pig iron produced during the preceding week was distributed among the four in proportions laid down in the agreement of partnership.[1] Similar product-sharing ventures existed in mining and the manufacture of glass. Generally, however, the partners in an enterprise subscribed to a common stock and took their profit in the form of money. Shipping was usually conducted in this way. A group of men pooled resources, each taking an eighth, a sixteenth, or a smaller fraction of the capital of a vessel, and each receiving his share of the profit (or contributing his share of the loss) at the end of the venture. So also in the mining industry of Cornwall. Local landowners, clergymen, craftsmen, and others joined with London or Bristol merchants or South Wales smelters. They entered their names in the 'cost book' of a mine, appointed a captain to supervise operations, and a purser to control finance, and then dispersed to meet again two or three months later to distribute gains or pay the calls needed to continue the undertaking. The wealthier adventurers spread their holdings over several widely scattered mines. They could give little time to the affairs of any one of these, and their

[1] A. Raistrick, *op. cit.*, p. 42.

Manufactures

status was essentially that of shareholders in a joint-stock. The venture itself, however, was a partnership. It had grown out of the simple arrangements of a group of working tinners; and the fact that the captain and purser were not only salaried officials but also partners, provided the incentive to efficiency that Adam Smith found to be lacking in companies.

The legal joint-stock concern, with corporate personality and rights to a monopoly, owed its existence to the grant of a royal charter. In the sixteenth and seventeenth centuries it had played a large part in overseas trade, as well as in mining and smelting. In the years following 1688, the ease with which fortunes could be made by buying shares and holding them for the rise had led to the formation of many unchartered, unincorporated companies. And after 1717 the capital gains made, or expected, by holders of South Sea stock led to another boom of flotations. At the height of this, in June 1720, a measure known as the Bubble Act was passed which prohibited, under very severe penalties, the creation of unincorporated companies. It has often been said that the Bubble Act impeded for more than a hundred years the rise of large-scale manufacturing businesses in England. Little evidence, however, has been offered. Only once, as far as is known, were proceedings taken under the Act, and this was soon after it appeared on the Statute Book—in 1724, in the case of a body inauspiciously named the North Sea Company. It was still possible to set up a joint stock by private Act: the canal companies were brought into being in this way. And the device of the equitable trust, under which mutual covenants were made between subscribers and trustees nominated by them, led to a growth of what were, in effect, companies in other fields of enterprise. All these had a joint stock, continuity of life, and transferable shares; and, late in the century, some of them found it possible to limit the liability of members.[1] If industry had really required a joint-stock organisation there seems to be little reason why it should not have had it. The few large concerns, such as the English Linen Company of 1764 and the Plate Glass Company of 1773, that were set up to control manufactures, had an unfortunate history.[2] Adam Smith was broadly right when he defined the field of profitable operations by joint-stock companies as that covered by banking, insurance, the provision of water, and the construction of canals.

---

[1] C. A. Cooke, *Corporation, Trust and Company*, ch. vi.
[2] D. Macpherson, *op. cit.*, vol. iii, pp. 401, 535.

The trust, on the other hand, had a somewhat wider sphere: it did much for the improvement of the roads, and it provided a legal basis for some manufacturing establishments, including the 'public' mills of the WestRiding which fulled cloth for both members and outside customers.

As already said, firms varied greatly in size. Where, for technical reasons, a large plant was required, and where the economies of integration were marked, the capital of an undertaking might run into scores, or even hundreds, of thousand pounds. Ambrose Crowley, who began as an apprentice to an ironmonger, in London, established a nail manufactory at Durham about 1684, and soon afterwards became the owner of slitting mills and steel furnaces at Winlaton, and of foundries and forges at Swalwell, where he made anchors, chains, and other heavy goods. He had large warehouses at London and Greenwich, and smaller depots at Blackwell, Ware, Wolverhampton, Walsall, and Stourbridge; and he owned a small fleet of vessels that plied between the Tyne and the Thames. In 1728, after the death of his son, John, the estate was valued at nearly £250,000. Not the whole of this consisted of industrial assets (Ambrose Crowley had been the largest shareholder in the South Sea Company), but it is clear that the business concern was by this time vast.[1] Later in the century the Walkers of Masboro', who had begun in a tiny smithy in 1741, built up a family business in ironsmelting the assets of which were valued in 1801 at £235,000. In addition, the associated firm of Walkers and Booth had a capital of £38,000, and the partners had also large holdings in lead works at Elswick, Derby, and London.[2] The Carron Iron Co. began in 1759 with a capital of £12,000 (divided into twenty-four shares): when it obtained a charter in 1773 its nominal capital was £150,000 (in 600 shares). In 1793 the firm of William Reynolds & Co. of Shropshire was valued at £138,000, and a few months later the Coalbrookdale Co., with which the Reynolds were associated, had an estimated capital of £62,500.[3] In the coal industry, pits were still relatively shallow and small, but the colliery embracing several pits was often a large undertaking, providing employment for 100 or 200 men. The copper mines of Cornwall, and the smelting works of

[1] M. W. Flinn, *The Law Book of the Crowley Ironworks*. Pub. Surtees Society, vol. CLXVII. See also Mr Flinn's *Men of Iron. The Crowleys in the Early Iron Industry* (1962).
[2] *Samuel Walker's Diary* (ed. A. H. John). Council for the Preservation of Business Archives, 1951.
[3] T. S. Ashton, *Iron and Steel in the Industrial Revolution*, pp. 43, 52.

Manufactures

South Wales had sometimes equally large capitals, but fewer employees; for their resources were in plant and stocks of ore or high-priced metal, rather than in funds for the payment of wages.[1] Paper-making, glass-making, and brewing also offer examples of big undertakings. After the death of Thrale in 1781, his brewery sold for £135,000[2]—and even at this price the purchasers were thought to have driven a hard bargain.

The organisation of industrial concerns was rarely complex. In a small business the proprietor supervised all operations. In a larger one each partner usually had his special duty: one saw to the purchase and delivery of materials, another directed the work, and a third kept the books and controlled sales. Ambrose Crowley built up a highly articulated structure: he appointed managers and supervisors to each of his establishments, drew up an elaborate code of 'laws', and directed not only policy, but day-to-day operations, by correspondence from London. In most concerns, however, the administration was rudimentary. The need to keep the workings within the bounds of the royalty, to direct the sinking of shafts, and to supervise sales, led the coal owners of the north to appoint a *bailiff of the field* at each colliery, as well as a *viewer* at each separate pit. But most of the work was done under contracts made with small companies of men, who undertook to sink a shaft or drive a level, at so much a yard, or by individual hewers, each of whom worked in his stall and himself paid the marrow who aided him. In the coal-fields of Lancashire, Yorkshire, and the Midlands, it was usual for a company of six or eight men to undertake to work an entire pi and deliver the coal to the owner, or his customer, at a price determined between the charter-master, or leader of the gang, and the proprietor.[3] Similar conditions prevailed in the copper-mines of Cornwall. The partners made contracts with *tutworkers* for the per formance of particular tasks, and with groups of *tributers* who agreed to work a pitch and to hand over a specified proportion of the ores they obtained.[4] Apart from controlling the winding-gear and the pumps, and sometimes providing candles and baskets, the proprietors had little concern with the working of the mines. In the smelting and refining of iron, in glass-making, and in most other

---

[1] For investment in smelting see G. Grant-Francis, *The Smelting of Copper in the Swansea District*, pp. 109, 115.
[2] Herbert Heaton in *Johnson's England*, vol. I, p. 252.
[3] T. S. Ashton and J. Sykes, *op. cit.*, ch. VII, *passim*.
[4] John Rowe, *op. cit.*, pp. 26-7.

manufactures there was the same devolution: even late in the century at the Soho Foundry of Boulton and Watt the fitting of nozzles was done by a group of men who worked for a collective piece-rate.[1] It was not the scale of operations, but the substitution of individual for group contracts, and the greater degree of supervision, that differentiated the factories from the earlier forms of manufacture—though even in the factories survivals of the older arrangements were common.

In his account of the royal dockyards Defoe had called special attention to 'the commissioners, clerks, accomptants, &c. within doors, the store-keepers, yard-keepers, dock-keepers, watchmen, and all other officers one to another respectively, as their degree and offices without doors, with the subordination of all officers one to another respectively, as their degree and offices require'. The growth of the size of private concerns called for a similar, if less elaborate, organisation: new classes of technicians, machine-makers, clerks, book-keepers, and supervisors appeared in the manufacturing areas. As, after the introduction of steam, the metal mines of Cornwall became deeper, the captains became salaried managers and the tributers foremen. In coal mining the charter-masters tended to become what are now called deputies. In the old-established trades of London it had been common for journeymen to undertake work by the piece and to employ other journeymen and apprentices as assistants: in the newer industries, such as brewing, distilling, sugar-refining, and soap-boiling, and wherever the scale of production was large, most of the workers were hired directly and supervised by foremen.[2] But the internal structure of the new factories and works, the relation between the numbers of skilled and unskilled workers, and the social consequences of the growth in scale, are matters that have received little serious investigation.

VI

A striking feature of economic life in the eighteenth century is the prevalence of combinations. There were, here and there, statutory bodies, like the Cutlers' Company of Hallamshire, and the Framework Knitters' Company, which exercised powers similar to those of the Stuart corporation; and in the metal-mining areas of Cornwall and Derbyshire, as well as in the Forest of Dean, even older

[1] E. Roll: *An Early Experiment in Industrial Organisation*, pp. 198, 200.
[2] M. D. George, *op. cit.*, p. 157.

Manufactures

regulative organisations persisted. Generally, however, it was less through powers conferred by government than by voluntary agreements that the industrialists exercised control. Dozens of instances are to hand, but a few will suffice for the present purpose.

In 1707, thirteen owners of fulling mills near Huddersfield agreed not to work on Sundays, not to deliver cloth until they had been paid, and not to charge less than stated fees for their work.[1] In 1712 the iron masters of Furness had an understanding as to the prices to be offered for charcoal or for coppices they might wish to buy outright, as well as to the quantity of fuel to be acquired by each concern. At the same time the iron smelters of Derbyshire and South Wales had arrangements for determining the prices of pig-iron.[2] In 1714 the associated copper smelters regulated the bidding for contracts with the Treasury; and in 1719 some of these united to control mines in Cornwall.[3] Later decades afford examples of powerful regional organisations. The colliery proprietors of the Tyne limited the quantity of coal shipped to London. They left mines unworked, rented staithes to prevent their being used, and bought land adjoining the river for the sole purpose of barring access to would-be competitors.[4] In 1751 the demand for vessels in the East Indian trade was below the supply, and the owners took concerted action to deal with the redundancy.[5] When the demand for atmospheric engines rose in the 'sixties Abraham Darby and John Wilkinson adopted a common policy for the sale of cylinders and pipes.[6] Smaller industrialists, such as the Sheffield steel tilters, silver-plate manufacturers, file smiths and tool makers, had agreements for fixing prices and terms of credit.

Towards the end of the century combinations on a national scale were by no means uncommon. In 1785 the Cornish Metal Company and the Anglesey Company divided the market for copper in agreed proportions. And at the same time the regional organisations of iron masters, potters, Birmingham manufacturers, and even the individualist cotton spinners, united in the General Chamber of Manufacturers to defend their common interests from the threat of taxation and competition from abroad. There must have been many hidden

[1] W. B. Crump and G. Ghorbal, *op. cit.*, p. 49.
[2] T. S. Ashton, *Iron and Steel in the Industrial Revolution*, p. 162.
[3] H. Hamilton, *op. cit.*, pp. 145-6.
[4] T. S. Ashton and J. Sykes, *op. cit.*, p. 212.
[5] L. S. Sutherland, *A London Merchant, 1695-1774*.
[6] T. S. Ashton, *op. cit.*, p. 164.

124   *An Economic History of England: the 18th Century*

understandings and agreements of which there is no record. But often there was no need for secrecy. The industrialists and the public had not yet absorbed the teaching of the economists as to the benefits of rivalry between producers, and there were not lacking voices to declare that competition must lead to a deterioration of quality.[1] Ordered trade, rather than increased output and lower prices, was still the aim of many who spoke with authority.

VII

The changes in organisation and technique led to a great expansion of output. There are no valid statistics of the production of coal, but the annual figures of exports from the great northern field may serve as a guide. For the decade 1701-10 they give an average of 183,000 Newcastle chaldrons; for 1791-1800 the figure is 758,000.[2] This fourfold increase is probably less than that of the inland coalfields, the output of which was not subject to duties and grew rapidly in the canal era. In 1731-40 the average annual production of copper in Cornwall was 7,500 tons; in the period 1794-1800 it was 48,000, and towards the end of the century almost equally large supplies were being mined in Anglesey.[3] Early statistics of the output of iron are unreliable. But about the year 1720 the production of pig iron can hardly have exceeded 25,000 tons: in 1788 the estimate was 61,000, in 1796, 109,000, and in 1806, 227,000 tons.[4] In the textile industries expansion was even more marked. The number of pieces of broad cloth milled in Yorkshire rose from an average of 34,400 in 1731-40, to one of 229,400 in 1791-1800. Between the first and the last decade of the century the annual output of printed cloths grew from 2·4 million to 25·9 million yards,[5] and the exports of cotton goods from a mere £24,000 in 1701 to £5,851,000 in 1800.[6]

---

[1] L. S. Sutherland, *op. cit.*, p. 109.

[2] T. S. Ashton and J. Sykes, *op. cit.*, Appendix E. The Newcastle chaldron weighed about 53 cwt.

[3] Figures for the output of copper in Cornwall are given for 1726 to 1775 by William Pryce, *Geologia Cornubiensis* (1778); those for later years are from Sir Charles Lemon, 'The Statistics of the Copper Mines of Cornwall', *Statistical Journal*, vol. I.

[4] T. S. Ashton, *Iron and Steel in the Industrial Revolution*, pp. 98, 235. The figures do not include the output of furnaces in Scotland.

[5] The figures for the output of broad cloth are from D. Macpherson, *Annals of Commerce*, vol. IV, pp. 15, 526; those for printed cloth from the Excise Returns.

[6] In 'official values'. The figure for 1701 is for England and Wales: that for 1800 relates to Great Britain.

Manufactures

When Arnold Toynbee gave currency to the term 'industrial revolution' he set the beginnings of the movement at 1760; and the tendency of later scholars has been to seek an earlier *terminus a quo*. The roots of modern industrial society can be traced back indefinitely into the past, and each historian is at liberty to select his own starting point. If, however, what is meant by the industrial revolution is a sudden quickening of the pace of output we must move the date forward, and not backward, from 1760. After 1782 almost every statistical series of production shows a sharp upward turn. More than half the growth in the shipments of coal and the mining of copper, more than three-quarters of the increase of broad cloths, four-fifths of that of printed cloth, and nine-tenths of the exports of cotton goods were concentrated in the last eighteen years of the century. After 1782 Englishmen turned, more than at any previous period, to the development of resources at home. If the secession of the American colonies led to despondency in political circles English manufacturers and inventors were pushing forward to new conquests. 'Three more balloons sail today', wrote Horace Walpole to Sir Horace Mann on May 13th, 1785, 'in short, we shall have a prodigious navy in the air, and then what signifies having lost the empire of the ocean?'[1] From the end of the war the number of patents leapt up: of the 26,000 belonging to the century considerably more than half were obtained after 1782.[2] A sudden increase of bankruptcies may not be indicative of prosperity, but a high average level, over a period of years, is the mark of an expanding economy. Of the bankruptcies registered between 1731 and 1800, about 44 per cent. fell in the years after 1782.[3]

The rapid development of English industry has been attributed to the exploitation of colonial peoples and to profits wrung from the slave trade. But it was after the Americans had won their independence, and at a time when the West Indian economy was in decline, that the pace quickened. It has been said that Britain built up her industrial power under the shelter of protective tariffs; and it is not to be denied that the woollen, cotton, and iron manufactures benefited from the prohibition of the export of wool, the ban on the import of printed calicoes, and the restrictions placed on industrial developments in Ireland and the colonies. But it was largely

[1] *The Letters of Horace Walpole*, vol. VIII, p. 552.
[2] Commissioners of Patents: *Titles of Patents of Invention, chronologically arranged by Bennet Woodcroft* (1854).
[3] The figures are taken from the monthly returns in *The Gentleman's Magazine*.

abroad that the expanding industries found their markets, and the efficacy of import duties in stimulating exports has yet to be demonstrated. Some of the duties were specific. As prices rose, after 1782, they bore a smaller and smaller proportion to the value of imported commodities: according to the estimate of Professor Imlah, in 1798-1800 customs duties amounted to only 18 per cent. of the real value of imports.[1] A tariff of this level can hardly explain the upward surge of output of manufacture.

Some writers, following Sombart, have laid stress on the part played by luxury and war in stimulating economic activity. The interest aroused by Lord Keynes in Mandeville's *Fable of the Bees* has tended to a belief that lavish spending, by individuals or the state, was a mainspring not only of a high level of employment, but also of industrial advance. If that were so we should expect development to have come first, and to have been most marked, in the industries that supplied the wants of the rich. In fact, it was primarily not in consumers' goods of any kind, but in capital goods: in the means of transport, and the output of coal, iron, copper, cotton and woollen yarn, and other semi-finished products—certainly not in that of jewellery, tapestries, laces, costly furniture, and porcelain. (This does not mean that neither rich nor poor benefited immediately from the expansion: a large part of the output was exported, and the things brought in return included tea, coffee, wines, sugar, grain, and other consumers' goods.)

War brought benefits to several industries. It was only when imports from the continent were cut off that the silk trade of Spitalfields was really prosperous. The demand of the armed forces stimulated the building of naval vessels and the production of iron, copper, lead, chemicals, leather, and the coarser kinds of woollen cloth. It induced some inventions, such as John Wilkinson's device for boring cannon. A high level of government expenditure tended to keep people in work. And, in an age when colonies were regarded as sources of supply of raw materials and as markets for the products of the mother country alone, the acquisition of new islands and plantations widened the field of trade.

But against all these have to be set, not only the loss of men and ships, but the derangement of peace-time economic life: the depression in the shipyards that built merchantmen; the decline in

[1] A. H. Imlah, 'Real Values in British Foreign Trade, 1798-1853', *Journal of Economic History*, vol. VIII, No. 2, pp. 133-52.

construction of houses, workshops, canals, and highways; the loss of markets in the enemy and neutral states of the continent, with its consequences for the makers of fine textiles; the depreciation of the currency; the increase of the national debt, and its effect on the distribution of income; the loss of technical training of the men serving with the forces; and (again if the figures of patents may be used as a criterion) the damping down of invention in general. The gains were perhaps more obvious, but this does not mean that they were more substantial, or more enduring, than the losses. The case for war as a stimulus to economic expansion is, to say the least, unproven. There is no need to look to extraneous forces for explanation of the growth of manufacture: it is to be found in the economic processes discussed above, which led to a lowering of costs and an extension of demand, not only by the rich and by the state, but by ordinary men and women.

A matter to which little attention has been paid is the extent to which the resources needed for the expansion of manufacture, transport, and building were obtained at the expense of other sections of the economy. Some capital, it is well known, came from abroad. In the early and middle years of the century a good deal (though less than has been believed) was imported from Holland.[1] But even before hostilities with the Dutch broke out in 1780, a part of this had been repatriated, and during the period of rapid industrial development there was a net efflux of funds to Holland. After 1789 well-to-do Frenchmen who feared expropriation, transmitted large sums (via Hamburg) to London. But though this refugee capital may have aided the expansion of credit in the early 'nineties, it can hardly have contributed much to the growth of manufacture. With the collapse of the assignats, and the reintroduction, in 1796, of a specie currency in France, most of it was withdrawn.[2]

Some of the labour required to sink mines, dig canals, and build factories and houses, was provided by immigrants. Enterprising Frenchmen played an important part in the textile industries, in particular; and there was a continuous flow of workers from Scotland, Ireland, and Wales. Since those who were prepared to face the hazards of a long journey were likely to be men of spirit and endurance they would have some, at least, of the qualities called

---

[1] Alice Carter, 'The Dutch and the English Public Debt in 1777', *Economica* (new series), vol. XX, pp. 159-61.
[2] R. G. Hawtrey, *Currency and Credit*, chs. XIII, XIV.

for. It is known from wages books that Welshmen helped to staff the ironworks of Shropshire and Staffordshire and the varied industrial establishments of Bristol and Liverpool. Scots supplied labour and enterprise to engineering in the midlands, and cotton spinning in Lancashire. And Ireland provided unskilled or semi-skilled men for the riverside trades of London, the textile industry of the northwest, and building and construction in all parts of the country. Though they sometimes flocked in when, as in 1782, there was famine in their native land, and when manufacture here was depressed,[1] more generally they came in response to a demand for labour in England. Their contribution to the growth of industry is not to be ignored.

In the main, however, England had to rely on her own resources. An economist might say that the capital required would be created by the act of investment itself: the ploughing back of profits by manufacturers, which was specially marked in this period, is one aspect of the process. But, as has already been pointed out, there were times at which the combined demands of government and industrialists drove up rates of interest to levels that impeded the development of agriculture. To some extent the increased labour required in manufacture, transport, and so on, was provided by the natural increase of population. But there was some deflection from other occupations. In the early part of the century a high proportion of the relatively poor was engaged in domestic service. Increasingly, well-to-do people complained that their personal attendants and chambermaids were being diverted from their proper sphere. (In what age have such complaints been unheard?) Landowners and farmers protested that workers were leaving the land. 'Let the country gentlemen be called forth and declare', wrote Corbyn Morris[2] about 1750, 'have they not continually felt, for many years past, an increasing want of husbandmen and day-labourers? Have the farmers throughout the kingdom no just complaints of the excessive increasing prices of workmen, and of the impossibility of procuring a sufficient number at any price?' There were similar protests in the building boom of the 'sixties: 'London now extends to Marylebone, to Tyburn, to Chelsea, to Brumpton, and some thousands of houses have been built within these last 3 or 4 years, which has

---

[1] A. Redford, *Labour Migration in England*, p. 116.
[2] Cited by George Chalmers, *The Comparative Strength of Great Britain* (1802), p. 121.

drained the country of all sorts of labourers and mechanics and raised wages.'[1] And towards the end of the canal boom, in 1793, the farmers of several counties asked for a suspension of work on the navigations until the harvests should have been got in.[2] The fact that, after the middle of the century, the prices of agricultural products rose more steeply than those of manufactured goods may perhaps also serve to support the assertion as to a deflection of workers.

There were, however, in England supplies of labour that had not been fully tapped. The early cotton factories were able to meet their wants by recruitment through the poor-law authorities. If the children had been left in the parishes in which they were born, some would have been apprenticed to trades and some put to domestic service. But others would have died or been sent overseas, and yet others would have grown up to be criminals, prostitutes, or permanent recipients of charity or poor-relief. Their employment in the cotton industry was hardly at the expense of production elsewhere. So, too, of large numbers of adults. Contemporary writers were unanimously of the belief that there were in parts of the countryside, and in some of the towns, more people than could find permanent employment there: a good many of the recruits to the new industries were squatters, semi-employed domestic workers, and paupers. It was the existence of such reserves of potential wage-earners that enabled England to acquire a transport system and a factory system, in a short span of time, without undue strain on other parts of her economy.

[1] *Considerations on the Scarcity of Corn and Provisions* (1767).
[2] *Sheffield Register*, February 8th, 1793. See also J. H. C., vol. XLVIII, pp. 273, 279, 305.

CHAPTER FIVE

Overseas Trade and Shipping

I

IN overseas commerce, as in domestic trade, there was a blending of old and new forms of organisation. From the Restoration, men of capital and enterprise had been knocking, with growing insistence, at the doors of the chartered companies. In 1673 the trade with Sweden, Norway, and Denmark was thrown open. In 1689 the Merchant Adventurers were shorn of most of their powers, and ordinary Englishmen became free to export cloth to all but certain reserved areas. In 1698 it was enacted that anyone might trade with Africa, subject to his contributing to the upkeep of the forts of the Royal Africa Company. And in the following year commerce with Russia and Newfoundland was declared open to all. On the other hand, some monopolies persisted. The Hudson's Bay and the Levant Companies had their privileges confirmed. The differences between the two groups that had contended for trade with the East were composed, and in 1709 the United East India Company came into being. This suffered, indeed, from the competition of interlopers, especially in its commerce with China; but it was not until as late as 1793 that outsiders were legally entitled to trade with the vast territories over which it exercised economic and political control.[1] The attempt made in 1710 to establish a similar monopoly in the trade of South America, however, came to nothing: the South Sea Company was, commercially, a failure from the start; and after the disasters of 1720, it became little more than a finance corporation with a special connection with the Treasury. Most commerce, except that with the East, was conducted by one-man concerns or family partnerships, and, though there were some strongholds of monopoly, most of the field lay open to competition.[2]

The qualifications needed for success as a merchant were set down in detail by Postlethwayt: some acquaintance with the commodities

---

[1] Even the ordinary trader had to send his goods in the Company's vessels.
[2] For details of the trading companies see E. Lipson, *The Economic History of England*, vol. II, ch. II.

of trade, command of foreign languages, and knowledge of book-keeping, shipping, foreign weights and measures, tariffs, foreign exchange, and the stock and produce markets. No single man could be an expert in all these matters, but he needed to know enough of each to be secure from deception or gross error. He might specialise in a particular commodity or a particular market, but it was of the essence of his calling that it transcended boundaries.

'It is not by the merchant as by the particular mechanic or artisan. [wrote Postlethwayt][1] The potter cannot easily strike into the business of the shipwright, any more than the latter can into that of the watch-maker or the weaver, &c. This is not parallel in regard to merchant and merchant. For the exporter of woollen goods, can as easily export tin or lead, or hardware, etc., and have his return by exchange, in dollars of Leghorn, or ducats of Venice, as well as in dollars of Spain, or millrees, or moidores of Portugal, &c. Or, cannot the merchant, who sent woollen goods to Spain, or to Italy, send another species of woollen goods to Russia; and have his return in robles, Russia hemp, linen, rhubarb, or pot-ashe &c. as easily as in Spanish dollars, wines, and raisins?'

Nevertheless, there was a growing tendency for men connected with trade to specialise. Some merchants, like John Pinney of Bristol and Edward Grace of London, had their own vessels, but increasingly shipping became a business distinct from trading. It was usual for a number of men to join to build and equip a ship, each taking one or more of the sixteen or thirty-two shares into which the capital was divided. The shipbuilder himself, the sail-maker, and the ship's chandler might take payment in the form of shares, and the master of the vessel, as well as some of those who intended to ship their goods by her, were also commonly among the partners. Often the master saw to the fitting, provisioning, and finance of the ship. But since he was away at sea for long periods it became usual to appoint some other partner to act as managing owner or, as he came to be called, ship's husband. The husband was often a man who, like Samuel Braund, had been a ship's master, and knew more about the sea than the rest of the partners.[2] Some husbands were polygamous: there is one instance of a man who controlled as many as twenty-four vessels. In the first half of the eighteenth century the ship's

[1] M. Postlethwayt, *The Merchant's Public Counting House* (1750), p. 31.
[2] Samuel was the brother of William Braund, the subject of L. S. Sutherland, *A London Merchant, 1695-1774*.

husbands became a clearly defined economic group: they flourished especially in the traffic in remote seas, and became powerful in the affairs of the East India Company. It was through them that the merchants chartered their vessels or hired space for their cargoes. But it was not always easy for a merchant to establish contact with the husband of a ship of the right kind and due to sail at the time he wished to send off his wares. Hence another specialised group arose—men known as ship's brokers, who, for a commission, were ready to meet the needs of the ship's husbands on the one hand and those of the merchants on the other. In the early part of the century they gathered at one or other of the coffee houses of the City—the Jerusalem, the Jamaica, and especially Sam's[1]—and after 1774 at the new Lloyd's coffee house in the Royal Exchange.

Similar developments took place in marine insurance. A rich merchant might be his own insurer: he might estimate the degree of risk to which his cargoes were exposed and rest content that the profit on those that arrived safely would cover the loss on those that went down at sea or fell into enemy hands. Sometimes owners of ships and merchants devised schemes for mutual insurance: this seems to have happened, in particular, in the shipping of coal. But, clearly, there was a place here for the specialist. The business of fire insurance had developed rapidly after the Great Fire of 1666; and in the first decade of the eighteenth century there was activity in the formation of life insurance companies. [2] Under the Bubble Act of 1720 joint-stock enterprise in marine insurance, however, was limited to two privileged undertakings, the London and the Royal Exchange Assurance companies. These incorporated bodies issued policies and made loans to merchants on bottomry and respondentia bonds (mortgages on hulls and cargoes), but they were unenterprising, and their premiums were high. Hence the bulk of marine insurance fell into the hands of outsiders. These were prohibited by statute not only from having a joint-stock, but even from forming partnerships; and thus it came about that underwriting took the form of individual enterprise.

Underwriting was no innovation: it had existed in Tudor times and perhaps earlier. But it was in the first half of the eighteenth century that, as a result of low rates of interest and a growing overseas trade, it became a specialised occupation. No individual could

---

[1] C. E. Fayle in *The Trade Winds* (ed. C. N. Parkinson), p. 30.
[2] It was not, however, until, with the creation of the Equitable Life Assurance Co. in 1762, that life insurance on a sound statistical basis began.

afford to insure, as the great companies might, the whole of a ship or its cargo. But there were men of moderate wealth who were ready, at any time, 'to write a line' for £100 or £500: acting co-operatively, they were generally able to offer lower rates and better credit than the joint-stock companies. But, here again, there was a place for intermediary agents to bring the shipowners and merchants into touch with the underwriters, and to see that the risk was adequately covered at a competitive price. These were the insurance brokers who came into prominence before the middle of the century and who, like the underwriters, met at the Coffee House of Edward Lloyd and, later, at its successor in the Royal Exchange. Here the underwriters sat each day in their boxes: the insurance brokers went from one to another, seeking the best terms for their clients, and adding name to name on the slip till the risk was covered. They saved the merchant the time and trouble of dealing with each separate underwriter; and for the traders or shipowners of the outports, in particular, they performed an essential service. Lloyd's was the centre of knowledge of everything relating to the sea. Here was the Register of Shipping (produced by a society of the larger underwriters) and here could be obtained the latest news, brought by packet boats or passing vessels, about the movement of ships between ports. Lloyd's, it was said, was better informed than Whitehall on what was happening abroad, on the prospects of war or peace, and the disposition of foreign navies.[1]

It was this information service that made it possible for underwriters to insure vessels that had already set sail and the position of which was unknown to their owners. Henry Hindley, of Mere in Wiltshire, sometimes omitted to insure his cargoes of linen yarn from Hamburg, and when these failed to arrive at the time expected he would send a hurried letter to his agent in London telling him to get them insured at once. There seems to have been no difficulty in carrying out such instructions. The men at Lloyd's would know that the vessel was held up by adverse winds off the mouth of the Thames or that the master had decided to run into Yarmouth or some other port. Needless to say, they were sometimes subjected to fraud. A ship's husband might over-insure his vessel and give the master instructions to run it ashore. And underwriters themselves might occasionally default. But Sir John Barnard, who underwrote

[1] For scholarly accounts of marine insurance see L. S. Sutherland, *op. cit.*, ch. III., and C. Wright and C. E. Fayle, *A History of Lloyd's*.

ships and cargoes, brushed aside this objection to his business by saying that 'if every part of commerce is to be prohibited, which has furnished villains with opportunities of deceit, we shall contract trade into a narrow compass'.[1]

A good deal of the business at Lloyd's was on behalf of foreigners. It is true that insurances could be taken out in other countries and that some of these had special advantages: each Spanish insurance company had its own patron saint, whose name appeared in every policy.[2] But generally rates were lower in London. Foreign merchants might instruct their English correspondents to insure vessels or cargoes for them, or might have direct dealings with a London insurance broker. Politicians and others disliked the giving of such aid to trade rivals: they objected to the practice of insuring neutral vessels in time of war against capture by the British Navy, and still more against that of insuring the ships of the enemy herself.

In 1741 there was a prolonged debate on the matter. Sir John Barnard pointed out that there were 'offices of insurance along the whole coast of the midland sea, among the Dutch and even among the French', and continued: 'That this trade is now carried out, chiefly by this nation, though not solely, is incontestable; but what can be inferred from that, but that we ought not to obstruct our own gain; that we ought not to make a law to deprive ourselves of that advantage, of which either favourable accidents or our own sagacity have put us in possession'.[3] If Englishmen refused to insure the ships of the Spaniards, the owners of these would go elsewhere. Perhaps members were impressed by the naïve declaration of Lord Baltimore that 'if our insurers gain by their practice, the Spaniards must undoubtedly be losers'.[4] But at an early stage of a greater struggle, in 1794, the insurance of enemy property was declared by the courts to be repugnant to public policy and consequently void.[5]

In addition to the husbands, ship-brokers, insurance brokers, and underwriters there were specialised agents and packers. To the merchants at the ports such men were useful: to those who carried on

[1] Cobbett's *Parliamentary History*, vol. XII (1741-3), p. 16.
[2] Joseph Townsend, *A Journey through Spain in the years 1786 and 1787* (1791). The author adds: 'Under this persuasion they ventured, about the year 1779, to insure the French West Indiamen at fifty per cent, when the English and the Dutch had refused to do it at any premium, and indeed when most of the ships were already in the English ports. By this fatal stroke all the insuring companies except two were ruined.' I am indebted to Professor G. P. Jones for this reference.
[3] *Application of the Merchants of London* (1742), p. 14.
[4] Cobbett's *Parliamentary History*, vol. III, p. 20.
[5] J. H. Clapham, *An Economic History of Modern Britain*, vol. I, p. 290.

their business from inland towns they were indispensable. Henry Hindley was an importer of linen yarn from Hamburg and Rotterdam and an exporter of cloths and linens to Ireland and Spain. That he was able to conduct his business from his home in Wiltshire was due to his being able to make use of the services of Timothy Budworth, a member of the Royal Exchange, who took charge of the yarn on the quay, paid the customs charges, and delivered the bundles to the carrier, and who also saw to the packing, shipping, and insuring of the cloths, and accepted and negotiated bills for Hindley. This was in 1763. Ten years later the same services were still being rendered for Hindley in London, but they were divided between three different men—evidence, perhaps, less of the growth of the merchant's business, than of a specialisation of the London agents.[1]

Sometimes merchants made use of supercargoes (often the masters of vessels). These voyaged with the goods, saw to their sale, and either remitted the proceeds, or, more often, bought foreign produce, made arrangements for shipping, and travelled home with the cargo. In the Virginia and Maryland trade the supercargo sometimes delivered the outgoing goods to dealers who had stores by the quays, and took tobacco in exchange;[2] in other cases he handed them to merchants to sell on commission. In trade with areas whose civilisations were different from those of Europe the supercargo played an essential part and fully earned his 5 per cent. fee. But elsewhere, once connections had been firmly established, it was possible to dispense with him and to send cargoes direct to a merchant, factor, or branch house in the foreign port. Factors were not an order of beings distinct from those who traded on their own account. Most merchants were willing either to buy and sell outright, or to obtain or execute orders in return for a commission. Merchants in all the great centres acted as factors for others: it was only when the volume of trade with any one of these places was large that it paid to set up a branch house for direct purchase and sale.

Yet other intermediaries who made straight the way of the merchants were those whose special concern was finance. Reference will be made below to the exchange brokers and bullion dealers of the City, some of whom towards the end of the century were

[1] *Hindley MSS.*, Wilts. Record Office.
[2] P. H. Giddens, 'Trade and Industry in Colonial Maryland, 1753-69', *Journal of Economic and Business History*, vol. IV, No. 3, p. 512.

evolving into merchant bankers. A little before 1750, Aaron Goldsmid, a native of Holland, set up in business in London as a dealer in foreign exchange. His eldest son, George, became a partner about 1773; and his second son, Asher, went into partnership with Abraham de Mottos Mocatta, bullion broker to the Bank of England. In 1776 two other sons, Benjamin and Abraham, established themselves as bill brokers, dealing first in foreign, and later in inland bills of exchange. In the 'eighties Benjamin married Jessie Soloman, the daughter of an East India merchant, who brought him a dowry of £100,000. The Goldsmids thus disposed of a large capital, and, with their wide international connections, were able to play a large part in the finance of English trade overseas.[1] Merchants needed not only short-term accommodation, but also facilities for the investment of balances arising out of their trade. The services rendered them by the stock-brokers of Jonathan's and Garroway's are too obvious to need further illustration.

It was natural that London merchants should have close associations with the state. Some of them, like Micajah Perry, acted as London agents to colonial governments, giving advice on such matters as trade regulation and currency.[2] In a century of wars in many parts of the earth, the British government had to make frequent calls on the knowledge of particular countries and the skill in transporting and remitting which only the merchant possessed. A man who, like Anthony Bacon, was able to obtain contracts for supplying the forces overseas might rise swiftly to fortune.[3] He was more likely to be successful in his application if he were himself a Member of Parliament: hence no small part of the resources of the larger London merchants went to the purchase of seats and to maintaining connections with the sources of patronage. Even more lucrative was the right to transmit money to allied monarchs or to the paymasters of the forces abroad. Mention is made elsewhere of the contracts held by Sir Henry Furnese, Peter Burrell, John Bristow and John Gore.[4] Other financiers, like Sir Joshua Vanneck, Samuel

---

[1] S. R. Cope, 'The Goldsmids and the Development of the London Money Market during the Napoleonic Wars', *Economica*, vol. IX (new ser.), No. 34, pp. 180-206. Dr. Cope disposes of the belief that Richardson, Overend & Co. were the first specialised bill brokers.

[2] Elizabeth Donnan, 'Eighteenth-Century English Merchants: Micajah Perry', *Journal of Economic and Business History*, vol. IV, No. 1 (1931), pp. 70-79.

[3] L. B. Namier, 'Anthony Bacon, M.P., an eighteenth-century merchant.' *Journal of Economic and Business History*, vol. II, No. 1, pp. 20-70.

[4] *Infra*, pp. 195-6.

Touchet, and Samson Gideon, were active in raising loans not only for British but also for foreign, allied, governments; and later in the century great houses, like the Barings, drew a major part of their incomes from similar activities. The extent of their trading and financial connections enabled them to underwrite large blocks of securities and to distribute the stock for which they had taken responsibility among the 'names' on their lists.

The close association between the government, the Bank, the South Sea Company, and the East India Company, was resented by the smaller and younger concerns in the City, and, under the vigorous leadership of Sir John Barnard, attempts were made to break into the circle. In 1750 Postlethwayt declared that 'the benefits of trading between the Exchange and the Exchequer' had been extraordinary, and that by means of it 'numbers acquired very great estates, without any great accomplishment'. He wanted to see more competition, and believed that the reduction of the rate of interest that followed the Peace of Aix-la-Chapelle would 'have a tendency to induce many persons of mature age into trade, as well as a greater number of young people to be trained up for it'.[1] Cheap money must have made it easier for parents to raise funds to pay the high apprenticeship fees demanded in London and to provide the liquid resources required by a youth entering into business on his own account. It must also have reduced the need for the state to make frequent appeals to the market, and so have had a stabilising influence on commerce. But before the final stage in Pelham's great conversion was reached, a new war had broken out. When Peter Burrell died in 1756 he was holding another contract for remitting to Jamaica, and shortly afterwards Sir George Colebrooke and Arnold Nesbitt had a covenant for the supply of money to North America. Men like Touchet and Vanneck did well out of their connections with the state during the Seven Years' War, though Touchet failed, disreputably, at the end of it.[2] Other instances could be drawn from the later wars of the century. It was partly through government contracts that the Thorntons, who had begun as Baltic merchants in Hull, were able to set up a large banking establishment (and also to give aid to Wilberforce in his agitation against the slave trade, and to distribute Bibles in all parts of the known world).[3]

[1] M. Postlethwayt, *op. cit.*, p. 13.
[2] A. P. Wadsworth and J. de L. Mann, *op. cit.*, pp. 247-8.
[3] F. A. Hayek, Introduction to Henry Thornton, *Enquiry into the Nature and Effects of the Paper Credit of Great Britain* (1802), pp. 13-14n.

And it was again partly, though by no means wholly, by the same means, that the Barings, who had entered on their English career in 1717, as linen and woollen merchants in Exeter, built up their great international house. Francis Baring had been trained by Samuel Touchet, who knew the ropes. By 1792 he was so well established in the City as to have become chairman of the East India Company, an office that brought him into intimate touch with government finance and, incidentally, enabled him to do service to English letters by finding a clerkship for Charles Lamb.[1]

In eighteenth-century parlance all the varied dealers mentioned above were merchants. The wide connotation was justified; for although some men followed a straight and narrow path others turned to diverse activities, without, however, relinquishing their business as merchants in the restricted sense of the word. Some of them had inherited the predatory and speculative proclivities of their Elizabethan forerunners. Sir Alexander Cairns engaged in a land scheme in Nova Scotia, and in 1709 was arranging to transport Palatines to Liverpool.[2] Sir Thomas Johnson, who was, more than any other merchant, responsible for building the dock at Liverpool, contracted to transport Jacobite rebels to the plantations, and sponsored a project to buy the French lands in St. Kitts.[3] And Samuel Touchet tried to get a grant of monopoly of the trade with Labrador, and fitted out, at his own cost, five vessels which seized Senegal from the French.[4] Other, more sober, merchants often transferred their activities from dealings in goods to finance. Between 1741 and 1774 William Braund was successively an exporter of woollens to Portugal, a shipowner, an importer of bullion, and an underwriter.[5] A few men, after having made a fortune in trade, turned their capital to manufacture: in his later years Anthony Bacon became one of the leading figures in the coal and iron industries of South Wales. But generally the progression was from trading in commodities to dealing in money and shares.

Some favoured youths were specially trained to be merchants. They received their first education in the elements of commerce in one or other of the academies (especially those of the dissenting bodies) which gave instruction in geography, modern languages, and

---

[1] R. W. Hidy, *The House of Baring in American Trade and Finance*, ch. I.
[2] A. P. Wadsworth and J. de L. Mann, *op. cit.*, p. 216n.
[3] C. N. Parkinson, *The Rise of the Port of Liverpool*, ch. 6.
[4] A. P. Wadsworth and J. de L. Mann, *op. cit.*, pp. 245-6.
[5] L. S. Sutherland, *op. cit., passim*.

book-keeping. After that they were apprenticed, at a cost of several hundred pounds, to a London merchant, and might remain in his counting house or be sent overseas. In the outports the fees were generally lower, and many boys who took early to the sea became merchants by stages, without any training other than in the school of experience. English merchants were drawn from all classes, except the poorest. Retailers like Stout pushed first into wholesale trade at home, then took shares in ships and sent cargoes across the seas. Supercargoes took to importing and exporting on their own account. Ship's masters became ship-brokers. John Julius Angerstein, a clerk in the office of a Muscovy merchant, came to be the largest underwriter in London. It is easy to see how a packer, like Thomas Burfoot, could graduate into commerce, or how a planter, like John Pinney, might return to Bristol and end his days as a merchant.[1] And it is equally easy to understand how manufacturers who produced goods of special quality (as in the case of John Huntsman, Jedediah Strutt, Josiah Wedgwood, Benjamin Gott, and Peter Stubs) should seek to deal direct with customers abroad. Some of these were willing to sell the goods of others (though not of competitors) on commission, and occasionally they assumed the title of merchant. Postlethwayt exhorted the gentry not to imagine that meddling in trade would 'corrupt their blood'. But it is doubtful whether his promptings were necessary. For many country gentlemen and men of title were of merchant stock, and some discovered that apprenticeship to merchant houses abroad might provide their sons with the education others purchased more dearly by the grand tour. In 1729 Elizabeth, widow of John Egerton of Tatton Park in Cheshire, bound her second son, John, to an English merchant in Venice, and, three years later, her third son to a merchant in Rotterdam. Both acquired knowledge of art, and both seem to have learnt how to strike a good bargain. There was a cosmopolitan atmosphere in the counting-house at Rotterdam, for, in addition to Thomas Egerton, the staff included a Swiss, an Irishman, a Frenchman, and a German.[2] If, in the eighteenth century, Englishmen became more conscious of their European heritage, it was through merchants, no less than scholars, that this came about.

To meet foreigners, however, it was not necessary to cross the

[1] Richard Pares, *A West India Fortune*, p. 163.
[2] W. H. Chaloner, 'The Egertons in Italy and the Netherlands, 1729-1734,' *Bulletin of the John Rylands Library*, vol. 32, pp. 157-170.

seas. On November 7th, 1760, a body of London merchants, led by John Gore, went to St. James's to condole with the new monarch on the death of his grandfather and to offer their congratulations on his accession to the throne. Some of them bore Anglo-Saxon or Welsh names that were, or were to become, famous: Hanway, Lloyd, Thornton, Walpole, Washington, Wilberforce, Wordsworth. But, alongside these, on the list, were others that cannot have come easily from English tongues: Vanneck, Muilman, Liebenrood, Boehm, Schoen, Schweighauser, Zurhorst, Motteaux, de Parthieu, de Visne, Fonnereau, Fernandes, Henriques, Espinoza, Montefiore, Pereire, Modigliani, and Ximines. Of the 810 merchants who kissed the hand of George III at least 250 must have been of alien origin.[1] It was one of the merits of the English at this time that they opened their doors to refugees and welcomed capital and enterprise from all quarters. Naturalisation was easy, and the strangers soon became good Englishmen: the absorptive capacity of Britain (like that of the United States at a later period) must be included among the influences that led to predominance in trade.

II

The greater part of British imports and exports was carried in British-built ships. Throughout the century the Thames estuary was by far the largest centre of construction: it was in the yards at Blackwall, Deptford, and Chatham that the naval and East India vessels were built by Barnard, Wells, Perry, and others[2]—though by the end of the period Portsmouth and Plymouth had become important in supplying ships for the navy. In the outports shipbuilding was a relatively small-scale industry, conducted by men dependent on timber merchants for their initial capital, and financed by mortgages on the ship under construction. But by 1740 John Okill & Co. of Liverpool were building fifth-rate and sixth-rate men-of-war of nearly 700 tons.[3] A substantial part of the mercantile marine was produced in the colonies, for the eastern seaboard of North America had ample supplies of timber, and the estuaries of the great rivers provided excellent sites for the purpose. The Americans developed high skill: their vessels seem generally to have been of better

---

[1] *London Gazette*, No. 10,050, November 4th-8th, 1760.
[2] L. S. Sutherland, *op. cit.*, p. 107.
[3] C. N. Parkinson, *The Rise of the Port of Liverpool*, pp. 105-6.

quality than those built in England. The secession of the thirteen colonies was a blow to British shipowners, but it came as a windfall to British builders. During the American war new yards were set up in remote creeks of Scotland and Wales and, for the first time, naval contracts were given to firms at Newcastle, Sunderland, and Leith.[1] No figures of annual output are available before 1787, when the Register Act of the previous year came into effect. The statistics relate to the British Empire, but now that the Americans stood outside this, nearly all the ships must have been built in Britain, and much the greater part in England. It seems probable that after the end of the war the yards had been busy in making up losses sustained by the merchantile marine, and that 1787 marked the culmination of a boom. Altogether 1,156 vessels, with a tonnage of 103,714, were built in this year. Thereafter the numbers fluctuate, falling to 714 vessels of an aggregate tonnage of 66,021 in the depression of 1794, and not until 1800 (when 965 vessels of an aggregate tonnage of 126,268 were built) was the figure of 1787 exceeded.[2] The relatively depressed state of shipbuilding during the war of the French Revolution was attributed to the use made of neutral vessels.[3] But it seems probable that the high level of interest rates was a contributory factor.

Until the middle of the seventeenth century no sharp distinction had existed between warship and trader.[4] Although after this time merchantmen continued to carry guns, and many served as privateers, it was only the large East India vessels that were thought of as belonging to the navy.[5] The merchant ships were of different types, each adapted to the seas in which they were to sail. It was, indeed, possible to transfer a vessel from one use to another. The slaver had to carry English merchandise to the Gold Coast, negroes to the West Indies, and sugar to England. After four (later six) voyages the ships of the East India Company were usually put into the African or West India service. On each of his three voyages of discovery Captain

---

[1] D. Macpherson, *op. cit.*, vol. III, p. 648; vol. IV, p. 10n. In Wales vessels were built at £6 10s. to £7 10s. a ton, a cost lower by 10s. to 30s. than those built on the Thames.

[2] The figures for 1787-92 are taken from *State of the Navigation, Commerce and Revenues of Great Britain* (MS. folio at the London School of Economics); those for the later years are from D. Macpherson, *op. cit.*, vol. IV.

[3] *Reports and Papers on Navigation and Trade* (1807), Introduction, pp. xx, xxi.

[4] G. N. Clark, *Guide to English Commercial Statistics, 1696-1782*, p. 46.

[5] R. G. Albion, *Forests and Sea Power*, p. 76.

Cook made use of a collier.[1] But generally the development of the shipbuilding industry exhibits a growing specialisation of product.

The growth of the merchant fleet and of its cargoes called for investment not only in the vessels themselves, but also in harbours, docks, quays, piers, warehouses, and other structures. London, it is true, carried out no great undertaking, but numerous schemes for improvements of harbours in the outports were set on foot, especially in periods of low rates of interest such as 1764-74 and 1785-93, as well as in the boom in overseas trade of 1799-1800. Though it was not impossible for a rich man to meet the cost of some of these from his own purse (Sir John Delaval made a new harbour at Hartley in 1761-4[2]), it was more usual to draw on collective resources. Individuals, municipal corporations, boards of trustees, and associations of merchants, all played a part. When, between 1710 and 1721, Liverpool created what was later known as its 'Old Dock', the initiative came from the corporation (leading members of which were shipowners and merchants), though London capitalists shared in the venture.[3] When, in 1774, a dry basin and new quays were constructed at Hull, the work was carried out by a company consisting of the corporation, the brotherhood of shipmasters, and several individuals. And when, in 1776, Bristol enlarged its floating dock and built a new quay and warehouses, it was the company of merchant venturers who were responsible. Where municipal organisation was weak, and voluntary association undeveloped, it was usual for the undertaking to be in the hands of persons nominated by Parliament. In some cases (as at Boston in 1776, Margate in 1787, and Neath in 1800) commissioners were appointed to act, either alone or in conjunction with the local corporation; and when a pier was built at the little port of Mevagissey in Cornwall the body responsible took the form of a trust.[4] Each undertaking had to be sanctioned by Parliament. Powers were given to recover the costs by levying tonnage duties, and occasionally financial aid was afforded: when the project for the new harbour was launched at Hull, the King gave a piece of land, and Parliament granted £15,000 out of the customs collected at the port.[5] Generally, however, the shipowners and

[1] J. A. Williamson in *Johnson's England* (ed. A. S. Turberville), vol. I, pp. 115, 120, 122.
[2] D. Macpherson, *op. cit.*, vol. III, p. 391.
[3] A. P. Wadsworth and J. de L. Mann, *op. cit.*, pp. 213-14.
[4] D. Macpherson, *op. cit.*, vol. III, pp. 554, 578, 587, vol. IV, p. 503.
[5] *Ibid.*, vol. III, p. 554.

Overseas Trade and Shipping

merchants supplied the capital. But when, in the opening years of the new century, London began to develop docks on a large scale, the money was raised from the general public: by this time investment in docks and quays, like that in shipping, had become a source of income for many who had no association with trade or the sea.

The growth of English shipping (as distinct from shipbuilding) owed little to scientific or technological developments. There was it is true, some progress in the arts of navigation and cartography. The invention of the chronometer by John Harrison in 1762 made it possible for the first time to determine with precision the degree of longitude at sea.[1] Improvements were made in rigging and sails. The use of copper for sheathing the bottom of ships, and the substitution of copper bolts for those of iron, lengthened the life of vessels in tropical seas.[2] And the device, hit upon in 1778, for 'reflecting the light of a great assemblage of lamps from about a thousand small mirrors' must have made the lives of mariners a little less hazardous.[3] But there were in shipping no such innovations as led to the transformation of the textile, iron, and chemical industries.

From the seventeenth century the commissioners of customs had maintained a general register of English and foreign vessels entering and clearing from English ports; and although most of the records have disappeared, summary tables are available for 1702 and for nearly every year from 1760 onwards. The figures must, however, be handled with caution. No account was taken of ships in ballast. Since imports consisted largely of raw materials, and exports of less bulky manufactured goods, departures in ballast normally exceeded arrivals in ballast, and for this reason the recorded figures of inward tonnage are almost always greater than those of outward tonnage. Some of the tables include all entrances and clearances, others take account only once of vessels that made repeated visits in a year. Moreover the figures of tonnage seem to have been very rough estimates, arrived at by multiplying the number of vessels by a conventional figure of tonnage which was higher for foreign than for English ships, and higher for London ships than for those of the outports.[4] In 1760 Macpherson said that the real tonnage might

---

[1] D. Macpherson, *op. cit.*, vol. III, p. 355.
[2] *Sailing Ships, their history and development*, Pt. I, p. 85. (Science Museum, 1935.)
[3] D. Macpherson, *op. cit.*, vol. III, p. 624.
[4] For an excellent account of the general registers of shipping and their deficiencies see G. N. Clark, *op. cit.*, pp. 45-51.

have been 'full fifty per cent above the reputed'.[1] In 1786, however, the Register Act sought to correct the disparity, and from the following year the figures are those of *ascertained* tonnages. A simple calculation (which need not be reproduced here) suggests that, at this time, the difference between ascertained and reputed tonnage was of the order of 15 per cent. But this does not mean that it would be safe to scale up the figures for all earlier years by this proportion.

Another set of figures of which little use has been made by modern scholars relates to the numbers of men serving on British vessels. In 1696 an Act was passed ordering that a deduction of sixpence a month should be made from the pay of all seamen, whether in English merchant ships or in the navy, for the support of Greenwich Hospital.[2] This was a burdensome tax for which only those who actually served with the fleet, and suffered in doing so, obtained any return; for the hospital was not open to ordinary merchant seamen. The historian cannot, however, regret the imposition. For the accounts of the Receiver of Sixpences from Merchant Seamen, which form a continuous series from 1718 to 1830, afford information as to the number of man-months served each year on British ships. They are the only employment statistics we have for any industry in the eighteenth century. Again, care must be taken in making use of them. From time to time regulations were altered and exemptions granted. But for such matters as the distribution of labour between London and the outports, and the short-term fluctuations in shipping, and hence in trade, they are invaluable.[3]

Enough has been said to indicate the nature of the statistical material available. The figures of ships relate to vessels of all kinds, from great East Indiamen to small coasters and tiny fishing smacks. Those of tonnage are the product of calculations made with the aid of a measuring rod that varied from place to place and time to time. And those of seamen, or rather of their sixpences, are open to other, though less serious, objections. But together with the figures of imports and exports (which will be discussed later), they suggest that, over the century, English activity on the sea increased roughly fourfold. They suggest also that changes took place in the distribution of shipping between ports. If we can trust the early returns, London was responsible for about a sixth of the ships, a third of

---

[1] D. Macpherson, *op. cit.*, vol. III, p. 340n. See also Ralph Davis, *The Rise of the English Shipping Industry* (1962). [2] 7 and 8 Will. III, c. 21.
[3] Ralph Davis, 'Seaman's Sixpences: An Index of Commercial Activity, 1697-1828', *Economica*, vol. XXIII, no. 2. (Nov. 1956).

Overseas Trade and Shipping

the tonnage, and rather more than a third of the seamen of England in 1702. At the end of the century, in 1800, she controlled a fifth of the ships and nearly two-fifths of the tonnage and men. (She was the port of the East Indiamen and other vessels of large tonnage and crews.) In 1702, according to the composite index, Bristol came first, Yarmouth second, and Liverpool third among the outports: by 1800 Liverpool had taken first place (evidence of the rise of the Lancashire cotton economy); Newcastle came second, and after that Hull and Sunderland. In vessels, tonnage, and men, Bristol was now eighth on the list (a result of the relative decline of the trade with the West Indies). The ranking of the ports towards the end of the century can also be seen in the figures of men requisitioned for the navy in 1795. London had to provide 5,704, Liverpool 1,711, Newcastle 1,240, Hull 731, Whitehaven 700, Sunderland 669 and Bristol 666.[1]

The change in the distribution of commerce, however, was not an unbroken process: in each period of war there was a tendency for shipping to move away from London to the outports, for enemy privateers naturally concentrated on the sea lanes from the capital port. 'It was observable', wrote William Stout, 'that the quantity of burthen of shipping rather increased at Liverpool, and the town increased in buildings and merchants during all the wars in King William's reign'; and there was a similar deflection of trade from the southern ports to Liverpool during the war of the Austrian Succession.

So far nothing has been said of ships other than those belonging to England and Wales. But a good deal of English and Welsh commerce was carried by vessels belonging to other parts of the British Isles, the colonies, and foreign nations. Under the Navigation Act of 1660 and the Staple Act of 1663, restrictions had been placed on alien-built or alien-owned ships. They were excluded from trade with the colonies, were not allowed to carry to England the products of Asia, Africa, or America, and had only limited rights of bringing in those of Europe. Some specified goods produced on the Continent (including timber, naval stores, and wine) were allowed in only if brought by British vessels, or by foreign vessels built in the country of origin or usual port of shipment of those goods. In all cases, commodities carried to England in foreign vessels were subject to an alien's duty.[2]

These restrictions had several objects. One was to wrest from the

---

[1] D. Macpherson, *op. cit.*, vol. II, p. 719; vol. IV, pp. 181, 340, 535.
[2] R. L. Schuyler, *The Fall of the Old Colonial System*, ch. I.

Dutch their primacy in the carrying trade. Another was to bind the colonies, by economic ties, to England. And yet another was to ensure to the navy a reserve of vessels that could be called upon in time of need. Perhaps it was the last of these that counted most in the eighteenth century. For trade itself, it was said, 'was considered principally as the means for promoting the employment of ships, and was encouraged chiefly as it conduced to the one great national object, the naval strength of the country'.[1] Judgment on the policy must depend on the weight attached to 'defence' on the one hand, and 'opulence' on the other. It is clear, however, that such success as it had was paid for by the British people. For, as Adam Smith wrote, 'if foreigners . . . are hindered from coming to sell, they cannot always afford to come to buy, because coming without a cargo, they must lose the freight from their own country to Great Britain. By diminishing the number of sellers, therefore, we necessarily diminish that of buyers, and are thus likely not only to buy foreign goods dearer, but to sell our own cheaper, than if there was a more perfect freedom of trade'.[2]

In each of the wars the British mercantile marine was depleted by capture on the seas. From 1702 to 1708 the number of sail taken by the French was 1,146; and from 1739 to 1742, it is said, 337 vessels were taken by the Spaniards. Between 1776 and February, 1778, 733 were captured by the rebellious colonists. And between 1793 and 1800 the losses by capture reached a total of 4,344, of which, however, 705 were recovered.[3] Again, in each of the wars, large numbers of English vessels were turned over to the navy for use as warships or carriers; others became privateers; and yet others that continued to trade were furnished with Letters of Marque which authorised them to capture any enemy vessel they might encounter.[4] Predatory activities, then, reduced the supply, and weakened the carrying capacity of the merchant marine. Freights rose sharply, and, with enemy privateers lurking outside English ports, rates of insurance were raised. Writing to order flax in July, 1761, Henry Hindley told his correspondent that it must be sent by a neutral ship, since the rate of insurance in this would be $1\frac{1}{2}$ per cent., as

---

[1] *Reports and Papers on Navigation and Trade* (1807), Introduction, p. xvii.
[2] *The Wealth of Nations*, vol. I, p. 408.
[3] *Reports and Papers on Navigation and Trade* (1807), p. clxxx. D. Macpherson, *op. cit.*, vol. III, p. 617. To a large extent the losses were made up by captures from the enemy.
[4] The difference between the privateer and the vessel with Letters of Marque is stressed by C. N. Parkinson, *The Rise of the Port of Liverpool*, p. 106.

against 6 per cent. in an English vessel. There were good reasons, therefore, for foreigners to extend their shipping within the limits allowed by the navigation laws. But, usually, in wartime, these laws were suspended and wider channels were opened to neutral vessels. At the height of the Seven Years' War, in 1761, no less than 26 per cent. of the tonnage entering British ports was foreign. When the war was over the proportion declined, and by 1775 it had fallen to 13 per cent. But the American War brought a reversal of the trend. In 1777 Macpherson pointed to the melancholy appearance of the Thames 'covered with foreign vessels, and particularly French ones, loading for various parts of the world with British cargoes, the shippers of which were now afraid to trust their property under the protection of the British flag'.[1] By 1782 36 per cent. of the tonnage entering British ports was foreign. The story was repeated in the last war of the century. In 1792 foreign ships accounted for less than 18 per cent. of the tonnage: in 1797 about 41 per cent. of the incoming tonnage belonged to neutrals—much of it to Americans. The position was naturally distasteful to British shipbuilders and shipowners:

'In consequence of the suspensions of the Act of Navigation since 1793, there has been a gradual decrease of shipbuilding in the British Empire, and a depreciation in the value of British shipping; those which were formerly employed in the circuitous trade from the Mother Country to America, and from thence to the British West India Settlements have been entirely driven out of it, which has since been carried on by Neutrals to the manifest injury of this Country. The facility with which Licences are granted to Neutrals affords a just and strong ground of complaint to British shipowners not only in the European Trade, which is principally carried on by them, but generally. . . .'[2]

No better evidence could be given of the success of the navigation policy. The navy had drawn on the mercantile marine, British traders on the reserve of foreign vessels. The desired effects of the laws (as was said of the Bank Charter Act of 1844) were experienced when they were in abeyance.

Policy was directed also to maintaining a large body of British-born seamen. There was rarely much difficulty in obtaining boys

---

[1] D. Macpherson, *op. cit.*, vol. III, pp. 352, 584, 611, 728; vol. IV, pp. 261, 439.
[2] *Reports on Navigation and Trade* (1807), pp. xx, xxi.

for the services. Adam Smith observed that 'without regarding the danger . . . young volunteers never enlist so readily as at the beginning of a new war; and though they have scarce any chance of preferment, they figure to themselves, in their youthful fancies, a thousand occasions of acquiring honour and distinction which never occur'. ('These romantic hopes', he added, 'make the whole price of their blood.') But this was of the soldier. 'The lottery of the sea', he continues, 'is not altogether so disadvantageous as that of the army. The son of a creditable labourer or artificer may frequently go to sea with his father's consent; but if he enlists as a soldier, it is always without it.'[1] Philanthropic institutions and Poor Law authorities apprenticed orphan boys to the sea, and when war broke out the watermen of the Thames, it was alleged, took steps to get their apprentices pressed into the navy so that they might draw the pay and prize-money.[2] But the state was not willing to trust to such influences alone. It was one thing to entice or impress boys, quite another matter to secure fully trained men. To have kept the naval establishment permanently at full strength would have been to waste human resources; for naval ratings, like soldiers, were 'as useless in times of peace as chimneys in summer'.[3] In 1696 an attempt was made to create a naval reserve. Thirty thousand mariners, fishermen, keelmen, and others were offered a bounty of 40s. a year each, on condition of their being ready, if called on, to man the fleet.[4] But after a few years this Register Act was repealed. At the end of the Seven Years' War another plan was put forward to keep in employment 250 warships, 1,500 officers, and 17,500 seamen, all of whom were to be paid by the state and employed in the whale fishery. They would be ready for service in any emergency, and the author of the proposal estimated that they might bring to the nation a profit of £100,000 a year. It was the kind of scheme that finds favour in the uncritical years of a post-war boom. But Macpherson questioned whether the additional number of whales caught would be in proportion to the additional ships, and, if it turned out to be so, whether the additional oil could be sold: 'The arithmetical rule of three is not in all cases infallible', he added.[5]

[1] *The Wealth of Nations*, vol. I, p. 97.
[2] M. D. George, *op. cit.*, pp. 142, 234, 383. In 1756 the Marine Society was founded to fit out poor men and boys for the navy.
[3] E. Wade, *Proposal for Improving and Adorning the Island of Britain* (1755), p. 18.
[4] D. Macpherson, *op. cit.*, vol. 2, p. 683.
[5] *Ibid.*, vol. III, pp. 368-9.

Overseas Trade and Shipping

The recruitment of able seamen to the navy in time of war was made more difficult by the rise of earnings in the merchant marine and by the chances of prize money offered by service in privateers.[1] It is true that the Lords of the Admiralty allowed some increase in pay in wartime, and, at the height of the struggle with France and Spain in 1758, an Act was passed to ensure that the men should be paid regularly, so that their wives and children should not be left (as they had been) in distress.[2] But such measures had little effect. Generally, reliance was placed on what a contemporary[3] called 'the barbarous and unconstitutional practice of pressing'. Naval vessels waited offshore to seize members of the crews of incoming or outgoing merchantmen, and press-gangs were active in the ports rounding up able-bodied men who might be turned into sailors.[4] Public authorities gave aid. In 1776, under order of the Lord Mayor, search was made of public houses in the city and many 'loose and disorderly men' were sent into the navy.[5] And in 1795 an Act was passed authorising the justices and magistrates of cities and towns to hand over for service with the fleet not only all rogues, vagabonds, smugglers, and embezzlers of naval stores, but also all other able-bodied, idle and disorderly persons exercising no lawful employment and not having 'some substance sufficient for their support and maintenance'.[6] But the press-gangs were not given to fine distinctions. Men might be abducted even when at work: in 1732 the artisans in a glass works at Whitefriars defended themselves by flinging molten glass at members of a gang who had forced their way in, and the officer in charge was almost scalded to death.[7] When, in 1756, James Watt was learning his trade as a mathematical instrument maker in London, he hardly stirred abroad, for he was neither an apprentice

---

[1] In the Seven Years' War the officers and crew of a privateer were allowed the whole value of the prizes taken. Similar opportunities of fortune were, of course, open to those in the navy. The capture of the *Hermione* (a Spanish ship laden with bullion and merchandise) by two British sloops of war brought to the admiral and the two captains £65,000 each: the commissioned officers had £13,000, the warrant officers about £4,335, and the men about £485 each. Ibid., vol. III, p. 357n.

[2] *Ibid.*, p. 309.

[3] Adam Anderson. D. Macpherson, *op. cit.*, vol. II, p. 683n.

[4] *The Application of the Merchants of London upon the Neglect of their Trade* (1742), pp. 14, 35.

[5] M. D. George, *op. cit.*, pp. 141-2, 362-3.

[6] D. Macpherson, *op. cit.*, vol. IV, pp. 340-1.

[7] *Whitehall Evening Post*, No. 2,208. I am indebted to Mr. L. M. Angus-Butterworth for this reference.

nor a tradesman and so would have had little hope of remedy if he had fallen into the hands of a press-gang.[1] And when in 1779, his assistant, William Murdoch, was on a journey from Birmingham to Cornwall, Watt sent letters to men of influence in Bristol and Exeter to solicit their protection for Murdoch.[2] Naturally, however, the danger was greatest for the sailor: in 1795, when each British port was called on to furnish its quota for the navy, one able seaman was accepted as equivalent to two ordinary men.[3]

The Navigation Acts required that three-quarters of the crew of an English vessel must be Englishmen. But in wartime this proportion could not easily be maintained. In 1740 it was enacted that so long as hostilities lasted a British merchant vessel might be navigated by any number of foreign seamen not exceeding three-fourths; and similar provision was made in each of the later wars of the century. As with ships, so with men, Britain was able to draw on a foreign reserve. What with alien vessels, and crews,[4] and British vessels also manned largely by aliens, the ports must have rung with strange tongues, and there must have been, at such times, an infusion of foreign blood. But, again, without this aid from neutrals, Britain could hardly have fought her wars and maintained her commerce.

III

The statistics of English overseas trade are the product of officials known as inspectors-general of imports and exports, the first of whom, William Culliford, was appointed in 1696. The ledgers kept by these men give details of the commodities of trade and specify the countries from and to which they were sent. In most cases particulars are given both of quantities and values. But since there is no way of equating hundredweights of coal and iron with pieces of cloth and pipes of wine except in terms of money, the total figures of trade with each country, and with the outside world as a whole, are necessarily expressed only in values. It was, indeed, the value of trade that interested contemporaries: the purpose for which the

[1] H. W. Dickinson and Rhys Jenkins, *James Watt and the Steam Engine*. p. 14.
[2] T. S. Ashton, *Iron and Steel in the Industrial Revolution*, p. 200n.
[3] D. Macpherson, vol. IV, p. 340.
[4] In 1792 the number of seamen on British ships entering English ports (including their repeated visits) was 81,901; that of seamen on foreign ships, 16,917. By 1800 the first of these figures had fallen to 62,676 and the second had risen to 37,803. D. Macpherson, *op. cit.*, vol. IV, pp. 261, 535.

inspector's office was established was to provide information as to the balance of payments with other nations.[1]

This object, it may be said at once, was never achieved. In most years the recorded value of exports was greater than that of imports; for the figures of exports were supposed to represent values of the goods when delivered at the foreign ports, including insurance and freight, whereas the figures of imports purported to represent values at the points of shipment abroad, and are therefore exclusive of these costs. In addition to the failure to take proper account of this, there are omissions of other kinds: the earnings of English fishermen on the banks of Newfoundland who sold their catches direct to foreigners; the profits of English ships carrying goods from one foreign port to another, or of English underwriters who insured foreign vessels; the incomes of English officials and agents abroad, and those of Irish and colonial landowners resident in England. These were 'invisible exports'. On the other hand, no account was taken of ships built in the plantations and sold to merchants and others resident in England, or of interest due to foreign holders of English stock. And no attempt was made to measure the movement of capital.

But, such matters apart, there are defects in the figures that make them of little use in answering questions less complicated than those of the balance of payments. The official values of imports into England and Wales were about £5·9 million in 1701 and about £28·4 million in 1800. The corresponding figures for exports were £6·9 million in 1701 and £40·8 million in 1800 (bullion and specie are excluded). If the figures could be accepted as accurate it would seem that imports had increased nearly fivefold, and exports nearly sixfold, and that the growth of trade over the century had therefore been somewhat greater than that of shipping. In fact, no such conclusion can be reached. There are grounds to believe that, after the removal of export duties in 1700 and 1721, merchants, for reasons of their own, tended to overstate the quantities of the things they exported, and that whenever an import duty was imposed or raised they tended to under-declare the quantities of goods they brought

---

[1] The *locus classicus* for the matters dealt with in the following paragraphs is G. N. Clark, *Guide to English Commercial Statistics, 1696-1782*. All students of English trade owe a debt of gratitude to Sir George Clark. I have not thought it necessary to make detailed acknowledgement for each statement taken from his pages: his index is comprehensive and the reader can find the source, without difficulty, himself.

in. This, however, is a minor matter. In determining the price to be set on each commodity in 1696 William Culliford consulted merchants engaged in overseas trade, and during the next year or two revised the valuations of some commodities so as to bring them into closer relation with current market prices. His successors, Charles Davenant and Henry Martin, seem occasionally to have followed the same practice; but by the end of the second decade of the century attempts at revision had virtually ceased. The valuations remained fixed; and whatever may have been the relation at the beginning, at the end of the century the official values were quite out of line with market prices. When the Convoy duty was imposed in 1798 merchants were required to declare the values of the goods they exported, and the able inspector-general, Thomas Irving, made a list showing, for a number of manufactured articles, the proportion by which the declared varied from the official values. In a few cases (beaver hats, linens, and sail-cloth) the official values were the higher, but generally the reverse was true. The declared value of unwrought copper exceeded the official value by 1,409 per cent. This was one of two extreme cases:[1] copper was important in warfare, and its market price was exceptionally high in 1798. But for flint glass the excess was 426, for coarse glass and earthenware 334, for tanned leather 283, for silk 113, and for wrought iron 103 per cent. For woollen goods, still the most important of British exports, it was 38 per cent.[2]

Almost from the start thoughtful people had been aware of the unsatisfactory character of the statistics of trade: most of the omissions referred to above, were mentioned by the early inspectors. But in 1718 Henry Martin found consolation in the following reflection:

'If the thing intended by these valuations was to discover at one view the increase or decrease of the quantitys of goods, imported or exported, it must be acknowledged that the keeping always to the same price of the same species of goods serves best for this purpose, since the increase or decrease of the quantitys must show at once in some measure this last increase or decrease: whereas if it had been possible to have brought into the total values the numberless variations of prices of all sorts of goods, this had not

---

[1] The other was haberdashery. Coffee was also grossly over-rated on exportation.
[2] D. Macpherson, *op. cit.*, vol. IV, p. 464. In 1800 the official value of the imports of Great Britain was £30·6 million, and that of the exports £43·2 million. Irving estimated that the 'real marketable value' of the imports was £55·4 million and that of the exports £55·8 million. (Ibid., vol. IV, p. 536.)

been sufficient to show the increase or decrease of the quantitys imported or exported, since it often happens that a less quantity of goods in one year is of more value than a greater of the same goods in another.'[1]

Other writers down to our own day have followed Martin in believing that, though the figures fail to represent values, they can be used as an index to the volume or quantity of trade. If we think of imports as divided into units, or (as Irving Fisher put it) baskets, each filled with duly weighted amounts of all commodities brought into the country, then changes in the official values of total imports represent changes in the number of such units or baskets. And so also for exports. But the validity of estimates of quantities of trade arrived at in this way surely depends on constancy of the unit. Not all commodities were valued at officially determined rates: against some entries in the ledgers are the words 'at value', which meant that the goods were rated according to the declaration of the merchant concerned. The value of exports entered in this way increased from under 3 per cent in 1700, to about 10 per cent of the total in 1796. The official prices of some exports varied with the destination of the goods, and a change in the direction of trade (e.g. from Portugal to the West Indies) might thus have the same effect on the figures as a change in volume. For these reasons, and because of the gradual introduction of new commodities into trade, the set of official values of the early years of the century became increasingly unsatisfactory as a yardstick. However, a test made by modern scholars,[2] who have applied instead the 'real' values obtained in 1798 to a number of years at intervals over the preceding century, suggests that no marked change in the picture would result from such a substitution. The conclusion reached is that, in spite of the deficiencies mentioned above, we are justified in regarding the official values as a rough index of changes in the volume of trade.

Between one year and another close in time it may be assumed, in the absence of evidence to the contrary, that any change in the constituents of trade was small. The statistics may thus be used as supplementary evidence of booms and slumps in overseas commerce. Moreover, they offer a mass of information as to the commodities dealt in and the rise and decline of particular markets. The few

[1] *Observations upon the Account of Exports and Imports for seventeen years.* Printed in G. N. Clark, *op. cit.*, p. 63.
[2] Phyllis Deane and W. A. Cole, *British Economic Growth 1688-1959*, pp. 43-4 and Appendix I (1962).

figures already given, and those that follow, are taken, by permission, from tables prepared by Mrs. Elizabeth B. Schumpeter, whose untimely death, in 1953, was a calamity to scholarship.[1]

Throughout the century, imports consisted largely of tropical and oriental produce, and raw materials. In the early years 'groceries', including sugar, tea, and spices, were much the largest group; but linens, wines, silk, tobacco, timber, and raw wool were also prominent. In 1800 groceries still held first place, and, after them, corn (it was a year of dearth), raw cotton, Indian piece goods, linens, flax, hemp, and dyestuffs. The raw materials of manufacture had increased in importance, and luxuries, like wines and tobacco, were now relatively low on the list.

By far the chief export was textiles. Among these, woollens and worsteds predominated: linens were far less important, and silks played an insignificant part. Outside this group the only important item in the first half of the century was corn, though the exports of fish, lead, and tin were by no means negligible. After 1750 exports of iron rose fairly quickly, and by 1800 were second (though a long second) to those of textiles; and refined sugar, of small account in 1701, was of some significance at the close of the century. The biggest change of all, however, was within the group of textiles: from tiny beginnings the exports of cotton goods had increased spectacularly, especially after 1782, and by the end of the period were, in official values, running very close to those of woollens.

Sources of supply and outlets for English wares changed from time to time. According to the estimates of Mrs. Schumpeter, at the beginning of the century about 53 per cent. of the imports of England and Wales came from Europe; fifty years later the proportion had fallen to 44, and by 1800 it was only 31. For exports the percentages at the three points of time were 78, 63, and 45.[2] The fall in relative importance does not represent an absolute decline of trade with the Continent: it was the result of a more rapid growth of commerce with other areas, especially with British overseas possessions, and, after 1783, with the United States.

English merchants had established connections with almost every European country: they brought timber, corn, iron, and naval stores

---

[1] With characteristic generosity, Mrs. Schumpeter put her tables, numbering about fifty, at my disposal. These and others have since been published in Elizabeth Boody Schumpeter, *English Overseas Trade Statistics, 1697–1808*.

[2] The percentages are for 1701-6, 1751-6, and 1796-1800. All include years of war, but the proportions are not very different for neighbouring five-year periods of peace.

from the Baltic, and silk, wine, and wheat from the Mediterranean. But the bulk of their trade was done with a few great ports: Hamburg, Amsterdam, Rotterdam, Antwerp, and (until the 'sixties) Lisbon and Oporto. These were points of assembly for goods from various parts of the Continent, as well as from overseas, and served also as distributing centres for English woollens and other manufactured wares. It was owing to this concentration that Henry Hindley of Mere was able to have dealings, at one remove, with spinners of flax in Germany,[1] and wholesalers who sold his black West of England cloth to peasants in Spain and Portugal. The Dutch and the Portuguese sent their own products to England—the Portuguese especially, since under the Methuen Treaty of 1703 their wines were admitted at rates of duty one-third lower than those levied on imports from France. But the exchange of native goods between England, on the one hand, and Holland and the Iberian peninsula on the other, was far less than might be assumed from the statistics. For the customs officials were concerned with the port of shipment and delivery, rather than with ultimate origin or destination; and so a good deal of trade with the Baltic appears in the returns as Dutch, and some of that with the Mediterranean and South America as Portuguese.

The commerce with France seems to have been small. This was partly because of national animosities (there were Anti-Gallican societies to discourage the drinking of claret and the wearing of French clothes) and partly because of preferential rates of duty on commodities from other sources. But a good deal of the exchange of merchandise between the two countries escaped the eye of the customs officials, not only because some of it was clandestine, but also because a large part was done through Holland and Flanders.

Apart from the European, as also from the colonial, was the Irish trade. Ireland has a soil and climate specially suited to pastoral and dairy farming. She supplied Britain and the colonies with cattle, meat, tallow, butter, and cheese, and took in return coal and manufactured goods from England, and flax seed from the colonies. She had formerly had large flocks of sheep and a moderately prosperous woollen industry. But, as a result of policies shaped in the late seventeenth centuries, the activities of her manufacturers were largely concentrated on linen. The industry developed mainly in Ulster,

[1] He had a special preference for the yarn of a spinner who is always referred to as 'the Widow'.

and after 1743 was stimulated by export bounties. In 1760 it was said[1] that 'the north of Ireland began to wear an aspect entirely new; and from being (through want of industry, business, and tillage) the almost exhausted nursery of our American plantations, soon became a populous scene of improvement, traffic, wealth and plenty; and is at this day a well planted district, considerable for numbers of well-affected, useful and industrious subjects'. There were, however, regions in which the subjects were not well-affected. The Irish were allowed to produce wool and woollen goods but might not export these to foreign countries; and generally their economic life was regulated so as to make it subserve the interests of England. Some of the restrictions were removed in 1778, when Ireland was allowed to trade direct with the Plantations and export woollen goods as she pleased. But others remained. Ireland had grievances which Englishmen ought not (and are unlikely to be allowed) to forget. Nevertheless, if her southern and western provinces failed to develop manufactures, the reason lay in their lack of coal and iron rather than in policy. And if, towards the end of the century especially, one of the chief exports of Ireland was men, this was due to the relative infertility of her soil and the high fertility of her people.

The trade between England and Ireland was important to both. According to the estimate of Lord Sheffield,[2] Irish imports increased threefold, and Irish exports nearly doubled, between the first and sixth decade of the century; and the progress continued. English imports from Ireland were greater than those from any country of Europe, and English exports to Ireland were almost as large as those to the East Indies.

Next in importance to the European and Irish commerce was that of the Atlantic. Here economies of diverse types were involved. There were the primitive, tribal communities of Africa, able to offer little more than ivory, gold dust, and slaves. There was the pioneer economy of the North American settlements, consisting of farmers, fishermen, and lumberers—again with relatively little to offer that Britain could not produce for herself. And further south, along the eastern seaboard and in the West Indies, there was the plantation economy, characterised by production of a single crop on large

---

[1] By a writer cited by D. Macpherson, *op. cit.*, vol. III, p. 318.
[2] *Observations on the Trade to Ireland*, p. 269. Cited by D. Macpherson, *op. cit.*, vol. III, p. 337.

estates by gangs of relatively unskilled workers under the direct or indirect supervision of resident owners. It supplied the sugar, tobacco, rice, cotton, indigo, and other tropical and semi-tropical produce in high demand in Britain and continental Europe.

English merchants, of London, Bristol, and Liverpool, had direct dealings, as exporters and importers, with each of these regions, but they also played a part, not fully represented in the statistics, in trade between one region and another. They sent gaily coloured Indian prints and hats, gems, flints, powder, knives, beads, and rum to West Africa, and exchanged these for elephants' teeth, gold, and, above all, negroes for the plantations. The plantation economy depended on the existence of sufficient land to enable the producer to move on when the soil became exhausted by monoculture. But it depended even more on a plentiful supply of docile workers. For, unlike wheat and barley, the plants called for detailed treatment, and there was little scope for the substitution of capital for labour. Attempts to make use of the native Indians had failed. Hence the demand for slaves.

The slave trade was not a creation of the white-skinned capitalist: it had existed time out of mind. But Europeans, and Englishmen in particular, carried it to a new height. Occasionally negroes were simply abducted; but pannyaring, as it was called, could never form the basis of permanent trade. Most of the men, women, and children were obtained by supplying English goods to native rulers or traders who brought the negroes, some of them taken in tribal wars, from the inland parts to the coast: often the forts maintained by the Royal Africa Company, or its successor, served as collecting centres. After examination by the ship's doctor, the negroes were taken on board and shipped to the West Indies or the American mainland. The merchants usually enjoined the masters of the vessels to care for the health of the negroes and to see that they suffered no maltreatment on the voyage: it was a matter of commercial prudence no less than of humanity. Nevertheless, especially in the early years, the rate of mortality on the seven weeks' passage was high, and sometimes the captives burst their fetters and murdered the crew.[1]

It has been said that the slave trade was highly lucrative, but it is

---

[1] Recent studies, based on the records of merchants engaged in the trade include J. A. Giuseppi, 'Pannyaring on the Coast', *The Old Lady of Threadneedle Street*, vol. xxvi, No. 17, pp. 320-4; C. N. Parkinson, *The Rise of the Port of Liverpool*, ch. 7; *Letters of a West Indian Trader* (ed. T. S. Ashton), Council for the Preservation of Business Archives.

not possible to disentangle the profits of slaving from those of the sale of English goods in Africa, or West Indian goods in England. The round trip was a single enterprise: to say that it must have been highly gainful because it yielded three profits is thoughtless. The crew of a slaver had to be larger and more skilled than that of an ordinary merchantman of comparable size, and the risks were high. Profits are a function of time. The round trip took nine to twelve months: in the same period more than three profits could have been made by trade with Europe. Often when the captain had set down his slaves in the West Indies he had to leave hurriedly to avoid the winter storms, and had no time to find a return cargo. In that case, payment for the slaves was made by bill of exchange, payable nine or twelve months ahead. Some Bristol and Liverpool merchants rose to fortune, but others lost their all: there is no reason to think that the trade afforded abnormally high returns.[1]

The settlers in North America disposed of large natural resources. But they too needed labour. Britain sent them convicts, prisoners of war, orphan children, and, increasingly, ordinary men and women who were willing to serve for a few years as indentured servants in the hope of setting up later as independent producers. She also sent capital—not only implements, ironware, nails and so on, but capital in the form of the consumers' goods which the colonists could not provide for themselves. As already remarked, Britain had no great need for the products of the northern colonies; but the West Indies required materials for building and fencing and food for the slaves. Lumber, staves, horses (for the crushing mills), grain, and fish were sent to them by the settlers of the American mainland, and in this way these met their adverse balance of trade with the mother country.

Exports to the British West Indies were of special consequence to the northern colonists, since their outlets for commerce elsewhere were limited by the Acts of Trade. They were obliged by law to obtain most of their imports from Britain; they were increasingly restricted from dealing with the foreign West Indies; and there was a growing list of 'enumerated' products which they were allowed to ship only to the mother country. Moreover, they were forbidden to manufacture cloth, hats, leatherware, refined iron and nails even for their own use. Like most thinly peopled communities, the colonies were short of metallic money, but the British government prohibited their

---

[1] F. E. Hyde, B. B. Parkinson and S. Marriner, 'The Nature and Profitability of the Liverpool Slave Trade', *Ec. Hist. Rev.*, sec. series, vol. V, No. 3, pp. 368-77.

issuing legal tender paper, and impeded their progress in other ways. It is true that the American colonists received something in return. They were free to admit immigrants from all countries. They were given aid against hostile Indians, and their vessels had the protection of the British navy. They had a privileged position in the markets of Britain and her empire, and were given subsidies on their production of naval stores and pig iron. And they were at liberty to sell unenumerated produce (including fish, livestock, grain and timber) to Europe. It is true, also, that the Acts of Trade were not fully enforced, that these were evaded by smuggling, and that, in spite of prohibitions, the colonists built up manufacturing industries. But they had real economic grievances. After the Seven Years' War new tensions arose: the British government was accused of favouring the East India and West India trade and the Hudson's Bay Company at the expense of the New England and Middle colonies. Attempts to make the colonists contribute to the cost of maintaining a standing army in America were resented. And there was antagonism to a monetary policy that was supposed to benefit the English creditor and injure the American debtor. Economic and political disabilities, combined, led to the revolt of the thirteen colonies and to the break up of the first British Empire.

After 1783 the ships of the United States were no longer able to trade directly with the British plantations, and the British West Indies suffered severely. But American goods were allowed into Britain at the same rates of duty as those levied on goods from British possessions, and the Atlantic trade as a whole expanded rapidly. In 1787-90 more than a third of the exports of the United States went to Britain, and no less than 87 per cent. of American imports of manufactured goods came from this country.[1] The Americans still took far more from Britain than they sent her, but they exported on a large scale to Ireland and continental Europe, and so enabled Britain to import more from these areas.

Another great region of trade was the East. India and China were thickly populated. They specialised in things, such as rice and silk, the production of which required much labour and little capital. The same was true of their manufactures of cotton and silk. Indian printed cottons made up a large part of the exports to England till 1721 when, in the interests of the woollen industry, their use was

---

[1] For further details see R. L. Schuyler, *The Fall of the Old Colonial System*, ch. III.

prohibited. The Chinese, it was said, were given to 'the sedentary arts of curious luxury', but it was not so much their pottery and lacquered work as their silk that found a market in England. Pepper and spices, indigo, cotton yarn, and saltpetre were among the commodities imported from the East. In spite of the pressure of West India traders, and in spite of the coffee houses popular among the merchants of London and Bristol, the English workers could not be induced to drink coffee. But they took with avidity to tea. Tea from China was by far the most important import from the East.

The export trade was conducted by the East India Company. It sought hard to induce the Indians and Chinese to protect their bodies from the tropical sun by means of English woollens. But both nature, and the religion of some of the people of India, were barriers. The Orientals were willing to take useful metals, such as copper, lead, tin, iron, and steel, and a considerable part of the exports of the Company consisted of these. But, most of all, the demand was for the precious metals, and especially for silver. The East India Company was subject to repeated attack on the ground that its operations were draining England of its currency (which was thought of as working capital). The defence it made was that by exporting silver it was able to bring back a large volume of goods that found a ready market on the Continent, and that the favourable balance of trade so established brought into the country more bullion than had been sent out. But no proof was ever given of this. With the East, as with the areas about the Atlantic, there was triangular as well as direct trade. After having unloaded their English cargoes in India, the ships carried raw cotton, pepper, ivory, sandalwood, and opium to the Hong merchants at Canton, who gave tea in exchange.

The Company had political as well as economic functions, and by the 'sixties had become 'a warlike, as well as a commercial commonwealth'.[1] Its servants were administrators no less than traders, and part of the adverse balance may be thought of as compensated by payments for services rendered in India. (It may also be thought of as loot.) There can be little doubt that, as a trading organisation, the Company had many defects. It guarded its monopoly carefully; and though it had to compete with Dutch and other companies, as well as with English interlopers, it managed to preserve most of its privileges unimpaired till 1813. It had many stations, from St. Helena

---

[1] D. Macpherson, *op. cit.*, vol. III, p. 387.

to Canton, and there was something in the argument that the maintenance of these, and still more, of ordered life in India, was a justification for its position of privilege. It is impossible to say whether, under an open trade, dealings with the East would have been greater or less. According to the highly fallible official statistics they accounted for less than 17 per cent. of English imports, and less than 6 per cent. of English exports, at the beginning of the century. By the end of it, when India had long ceased to send her fine fabrics, and when, indeed, English plain and printed cottons were beginning to find a market in Bengal, the share of the East in imports to England was still only about 20, and that in the exports about 10 per cent.

A large part of the produce brought from both western and eastern seas was not for British consumption. A nation with a large merchant fleet and colonial possessions almost inevitably develops an entrepôt trade. Even in the seventeenth century a good deal of the tropical produce brought to England was re-exported to the Continent; and it became a major object of policy to make Britain 'the common depositum, magazine, or storehouse for Europe and America, so that the medium profit might be made to centre here'.[1] It would be wrong to think of this profit as easily made: the real costs of collecting, transporting, warehousing, and selling to foreigners were by no means small. Risks were high. Only a nation that had low rates of interest and insurance could have hoped to carry on an extensive intermediary trade. The statistics are too unreliable to make it possible to give even approximate estimates of the value (or 'volume') of re-exports. On entry to this country tobacco was rated at 19s. a hundredweight: on re-export its rating was 38s.; and there was an equally wide discrepancy between the import and re-export valuations of coffee (a commodity that, towards the end of the century, played a leading part in the entrepôt trade). In official values, re-exports were generally about a third of total exports. Whether in terms of market prices they were above or below this, it is impossible to say.

In the first two decades of the century, about four-fifths of the imports and two-thirds of the exports of England passed through the metropolis. London was by far the largest centre of production

---

[1] J. Tucker, *Instructions to Travellers*, p. 13. Not all the re-exports had to be stored. Goods might be unloaded, duties paid, goods put on board again, and drawbacks received, within a few hours. D. Defoe, *op. cit.*, vol. I, p. 122.

and consumption. She was better situated than the outports for trade with the Continent; the chartered companies had their headquarters here; and the larger shipowners and the underwriters did their business in the City. London merchants had the ear of the government: they subscribed to national loans and could expect favours in return. (When in 1750 the plantations were permitted to make bar iron they were, for some years, allowed to export it only to London; and the same was true of French lawns and cambrics after 1767.) The entrepôt trade was focused on the capital: London alone had the wharves and warehouses required for the enterprise. Gradually, however, the concentration of trade on the metropolis was reduced. New industrial centres grew up in the north; new canals connected these with the outports; and manufacturers were increasingly disposed to market their output through Liverpool, Newcastle, and Hull. Trade with the West Indies and America increased more rapidly than that with Europe, and Liverpool and Bristol were more favourably placed than London for this. In each period of war ships were deflected from the Thames to estuaries off which the dangers of capture by privateers was less; and some of them, no doubt, continued to use these in times of peace. Moreover, by the middle 'eighties if not earlier, the organised cotton merchants and manufacturers, the potters, the iron producers, and the hardware dealers had learnt something of the art of lobbying: it is possible that the influence of the City on government was weakening. Whereas at the beginning of the century only a fifth of the imports into England and Wales came to the outports, by the end of it their share had risen to a third. For exports the change was less marked. At the beginning of the century a little over 33 per cent. had passed through the outports: in 1800 the percentage was 38. London still engrossed most of the entrepôt trade: for British exports alone the percentage belonging to the outports was 41.[1]

A good deal of exporting and importing took place outside the authorised channels of trade. This was the result of a system of protective and revenue tariffs only the barest outline of which can be given here. From early in the seventeenth century the export of wool had been prohibited, and until the reforms of Walpole there was a general export duty. In 1698 the standard rate of tax on imports was 10 per cent., but, under the pressure of wars, this was increased to 15 per cent. in 1704, 20 per cent. in 1747, and 25 per

[1] D. Macpherson, *op. cit.*, vol. IV, p. 536.

cent. in 1759. During the American War of Independence, in 1779 and 1782, further increases were made, each of 5 per cent. on the existing duties. In addition, there were special taxes on the import of a number of commodities the demand for which was thought to be highly inelastic, and, from time to time (notably in 1796), increases were made in these. In spite of the reforms of Pitt in the 'eighties, the level of duties was higher at the end than at the beginning of the century.

Prohibitions and high duties gave rise to smuggling. The profits of the clandestine trade depended on the existence of a marked difference between prices in England and those abroad. In the case of certain kinds of wool, prices were much higher in France, and the temptation to export these was great. In 1744 George Bridges, who, after nine years as a smuggler had found a more respectable, if less respected, occupation as 'a destroyer of buggs', described some of the ways in which the trade was conducted.[1] First of all, the Irish, who were supposed to send their wool only to England, shipped a good deal of it to the Continent. English graziers delivered supplies, openly, to Scotland and the Orkneys, whence the wool had an easy passage to Holland, Flanders, or France. Combed wools, as well as shearings, were sent by road to 'men of fortune' who owned houses or cellars at points on the south coast of England, and had specially built craft to carry the contraband to ships lying offshore. Other supplies were loaded into barges on rivers and floated down the estuaries to points at which vessels that had already cleared the Customs were waiting by night to receive them. Some wool and yarn was pressed into bales and passed through the Customs as drapery, and some went abroad, tightly packed in the baggage of passengers.

The commodities subject to special duties on importation naturally sold at higher prices in England than abroad. Most of them, like tobacco, wines, spirits, tea, lace, silk, and printed calicoes[2] were of small weight and bulk in relation to their value and lent themselves readily to smuggling. In the case of tea the incentive to smuggle was especially great, for legal imports were controlled by the East India Company which, following the usual practice of monopolists, limited sales and maintained prices at a level far above that of

[1] George Bridges, *Plain Dealing, or the whole Method of Wool Smuggling clearly Discovered* (1744).
[2] There is evidence of a good deal of smuggling of printed calicoes both before and after the Act of 1721. A. P. Wadsworth and J. de L. Mann, *op. cit.*, p. 139.

Hamburg and Amsterdam. Smuggling flourished especially in remote or isolated parts of the British Isles: it was carried on along the whole coast, but was particularly concentrated in the ancient liberty of Romney Marsh and the Duchy of Cornwall. Until 1765, when the Duke of Athol sold his sovereignty to the Crown, the Isle of Man was a nest of smugglers; and throughout the century Ireland and the Channel Isles were main centres of the business.

The lines of demarcation between pilferage from vessels in port, smuggling, privateering, and piracy are not easily drawn. There was much smuggling by the watermen of the Thames, and, at the end of each of the wars, privateering vessels took to running in illicit cargoes or acted as convoys for the luggers engaged in the trade. They were well armed and manned by resolute scoundrels, and the revenue cutters often thought it well to avoid them. Some of them made regular sailings between England and France, but, perhaps more frequently, cargoes were picked up at sea from homeward-bound vessels. Fishing smacks, packet boats, and occasionally even the revenue cutters themselves, took a part in the traffic.

A good deal of tobacco was brought in by seamen who distributed small bundles about their persons: hence it was decreed that no tobacco should be imported from the Plantations, or shipped there, except in casks or chests of at least two hundredweight.[1] Much of the tobacco (as well as wines and tea) brought in consisted of supplies that had already passed through the Customs in the ordinary way and had been allowed a drawback on re-exportation. They were taken up at sea from the outgoing ships and unloaded by night at various parts of the coast. In London smuggled tobacco was sold by the coffee houses, and in the country by hawkers and pedlars. Often duties were evaded by false declaration and intermixture of goods: in the Channel Islands French wines were added to those brought from Lisbon, and the whole was passed through the Customs on payment of the lower duty levied on port. There was hardly any ingenious device known to later generations that had not been employed in the eighteenth century.

The measures taken to suppress smuggling seem to have been of little effect. In 1698 some 300 riding officers had been appointed to check the export of wool and the import of uncustomed goods.[2]

---

[1] Alfred Rive, 'A Short History of Tobacco Smuggling', *Economic History*, No. 4 (1929), p. 558. Similarly, spirits had to be imported in casks of at least 60 gallons.
[2] Alfred Rive, *loc. cit.*, p. 560.

But these had been given an impossible task. They were terrorised by the well-organised, armed gangs of smugglers: some of them not merely connived at, but took part in the trade, and when high rewards were offered for seizures, they shared these with the smugglers.

As Steele remarked, it is of the nature of mankind to love everything that is prohibited. Most members of the public looked with favour on those who provided them with cheap spirits, tobacco, and tea; and even philosophers were on the side of the law-breakers. In a well-known passage Adam Smith described the smuggler as 'a person who, though no doubt highly blameable for violating the laws of his country, is frequently incapable of violating those of natural justice, and would have been, in every respect, an excellent citizen had not the laws of his country made that a crime which nature never meant to be so'. He was scornful of those who took a high moral line: 'To pretend to have any scruple about buying smuggled goods would in most countries be regarded as one of those pedantic pieces of hypocrisy which, instead of gaining credit with anybody, serve only to expose the person who affects to practise them to being a greater knave than most of his neighbours.'[1] It was a tolerant and cynical age.

The only possible cure for smuggling was to remove the incentive. In 1784, the accountant of the East India Company, who had good reason to know about such matters, estimated that, of the tea consumed in Britain, hardly a third had been legally imported: the rest had come in surreptitiously, mainly from the Continent.[2] It was believed that the illicit trade in tea encouraged the smuggling of other things that would not have been brought in 'except, as it were, in the train of a more capital, or more convenient article'. Hence, as a measure to reduce smuggling in general, it was decided to lower the import duty on tea from the 119 per cent. at which it stood to $12\frac{1}{2}$ per cent. Whatever the effect on the contraband trade in other commodities, the volume of tea that passed through the customs house increased vastly: in 1784 the amount entered for home consumption was 4,962,000 lb.; in the following year it was 16,307,000 lb.[3] Rarely can the accuracy of a statistical estimate have been more decisively confirmed: the accountant must have been delighted.[4]

[1] *The Wealth of Nations*, vol. II, pp. 379-80.
[2] D. Macpherson, *loc. cit.*, vol. IV, p. 49n.   [3] *Ibid.*, p. 336.
[4] He had estimated that the quantity of tea smuggled was 12,258,000 lb. It should, however, be remembered that the legal market for tea was controlled on the side of supply, and that the East India Company had a case to establish.

The period in which this experiment was made saw other measures to remove barriers to trade. English manufacturers, with their new techniques, had little to fear from foreign competition and were eager to extend their markets. The attempt to arrange a treaty of commerce with Ireland was, it is true, abortive. But the Eden Treaty with France in 1786 raised high hopes, and brought material gains to both parties. And in 1792 the East India trade was thrown open. At the same time an attempt was made to mitigate the worst horrors of the traffic in slaves: in 1788 limits were set to the numbers to be carried in vessels of prescribed tonnage; and there were prospects of a speedy end to the whole trade. But the outbreak of war in 1793 brought reaction. The necessities of the Treasury led to successive increases of duties, and liberal causes were frowned on. Neither increased taxation, nor the revival of smuggling that went with it, was able, however, to arrest the upward movement of legitimate trade that had begun in the 'eighties. Industrial output was increasing too rapidly for that. One is left wondering what peaks might have been reached, both in commerce and human liberty, if the war could have been avoided. But, one is told, speculation of this kind is unbecoming to an historian.

CHAPTER SIX

Money, Banking and Foreign Exchange

I

IN the eighteenth century the unit of account in England was (as it is today) the pound sterling. From the time of Queen Elizabeth this had been identified with a fixed quantity of silver, which was hence said to be the standard of value. The coins in circulation consisted of a variety of pieces of money, made of silver, gold or copper, which were supplied without charge to those who presented bullion for coining. In spite, however, of the absence of seigniorage the supply of legal money was far from adequate to the needs of an expanding economy, and much inconvenience and social disharmony arose from a shortage of coins of a kind suitable for the payment of wages and for retail transactions.

For this state of affairs part of the responsibility lay with the Royal Mint, a body which, though answerable to the Treasury, preserved many of the features of the privileged corporation of Stuart times. At the head was an official, known as the Master and Chief Worker, who exercised loose control over a number of specialists (the Assayer, Melter, Refiner, Engraver, and Medallist) and entered into contracts with a Company of Moneyers which was responsible for the manufacture of the coins. Until as late as 1799 the Master of the Mint received his remuneration in the form of a commission on output. Each of the specialists was allowed to combine the execution of his public duties with work on his own account. And although the individual moneyer was guaranteed a small, fixed annual payment 'lest he be too exposed to temptation', he drew the bulk of his income from the profits of the contracts made with the Master. Membership of the Company was restricted by the imposition of high fees for apprenticeship: about the middle of the seventeenth century there had been fifty-nine moneyers, but by 1705 the number had been reduced to sixteen, and by 1774 to eight (assisted by four apprentices). The large profits on the recoinage of gold in 1773-4, it is said, raised the status of the moneyers well above that of day labourers or artisans, and enabled the Company to invest in real estate on a considerable scale. In a petition to the Treasury in 1806 the moneyers declared that they 'would not have been able to support themselves at all, much less to hold that respectable rank in society commensurate with the situation of high trust in which they are placed, had they not possessed some patrimonial property of their own'; and by the middle of the nineteenth century the position of privilege enjoyed by these men had become little less than a scandal.[1]

The fact that the senior officers of the Mint had the right to work on their own account meant that orders from private individuals often took precedence over those from the Treasury. And since, until 1770, payment for coining was proportioned not to the number of pieces of money struck, but to their aggregate value, it was in the interests of the moneyers to coin gold rather than silver, and pieces of large, rather than of small, denomination.[2]

Even, however, if the Mint had been well conducted, monetary arrangements in Britain and abroad were such as to lead unavoidably to a shortage of silver coin. Both gold and silver were brought from the mines of South America to Spain and Portugal and distributed to all parts of Europe, according to demands exercised through international trade. In England silver coins were legal tender for any amount; gold pieces, however, had only a limited tender, and it was customary to express their value in terms of silver shillings. When the guinea was first issued in 1663, it was left to find its own price on the market, but within a short time it was found expedient for the state to declare that it would accept it, in payment of taxes, as equivalent to a fixed number of shillings.

In the early years of the eighteenth century the rate was 21s. 6d., and this came to be the price at which the guinea changed hands in commercial, as well as fiscal, transactions. At this reckoning it took a little more than $15\frac{1}{2}$ pounds of fine silver in the form of coins to buy one pound of fine gold in the same form:[3] in other words,

[1] *Report on the Constitution, Management and Expense of the Royal Mint*, B.P.P. (1849).
[2] Except when, as in the case of the five-shilling piece, the production of the larger coin required more labour.
[3] According to the Mint indentures, a pound (Troy) of gold, 11/12ths fine, was made into $44\frac{1}{2}$ guineas: hence a pound of pure gold was the equivalent of 48 6/11ths guineas. Similarly, a pound (Troy) of silver, 111/120ths fine, was made into 62 shillings: hence a pound of pure silver was the equivalent of 67 3/111th shillings. When the guinea was rated at 21s. 6d., 48 6/11th guineas were worth 1,043 8/11th shillings. Hence a pound of pure gold in coin was worth 1,043 8/11 ÷ 67 3/111 = 15·57 pounds of pure silver.

the Mint ratio was $15\frac{1}{2}$ to 1. In Spain and Portugal currency arrangements were such as to produce a ratio of about 16 to 1, and since 'the merchant will always make that metal his standard which is highest valued at the mint', the traders of these nations tended to use gold for internal transactions and to export silver to England where its value in terms of gold was higher. In most other Continental countries, however, gold was rated by the currency authorities at only about fifteen times its weight in silver: it therefore paid merchants in these countries to export gold to England and take silver in exchange. More important was the fact that silver changed hands for gold in India at a ratio of about 12 to 1 (and in Japan at a ratio as low as 9 or 10 to 1). In these circumstances there was a substantial profit to be made by shipping silver from England (and other countries of Europe) to the East. Under English law it was permissible to export the precious metals in the form of bullion, or of pieces of foreign money, but illegal to melt down coin of the realm for this purpose. Hence among dealers a pound of silver bullion would buy more gold than would a pound of silver coin: there was a divergence between the 'market ratio' and the 'mint ratio'. A writer[1] of 1762 put the matter quite simply in the following words:

'By law, 62 shillings are to be coined out of One pound, or 12 Ounces of Standard Silver.—This is 62 pence an Ounce. Melt these 62 shillings, and in a Bar this Pound Weight *at Market*, will fetch 68 pence an ounce, or 68 Shillings. The Difference therefore between coined and uncoined Silver in *Great Britain is now* 9 2/3 per Cent.'

Such an opportunity of gain was not neglected. Money jobbers and exchange dealers ('and herein', as Joseph Harris[2] averred, 'the uncircumcised' were 'just as good marksmen as the sons of circumcision') picked out the heavier of the silver pieces, threw them into the crucible, and then gave their oath that the resulting ingots were the product of English plate or of foreign silver coin. It was only the old worn or clipped pieces that were left in circulation: the rest had gone to grace the bodies of women in India, to provide votive offerings in the temples of China, or simply to swell hoards in these far-off places.

[1] The author of *Reflections on Coin in General*. See J. R. McCulloch *a Select Collection of scarce and valuable tracts on Money* (1856, reissued 1933), p. 519.
[2] *Essay upon Money and Coins*, McCulloch, *loc. cit.*, pp. 463-4.

During the great recoinage of 1696-8 most of the old clipped or worn silver pieces had been called in and replaced by new full-weight coins with milled edges. It was disturbing to the currency authorities to find that, almost immediately, large quantities of the new money disappeared from circulation, and that relatively little silver bullion was being brought to the Mint. In 1701 Isaac Newton pointed out that the prohibition of the export of English coin discouraged the import of silver 'because the merchant can make no use of it whilst it stays here in the form of bullion'.[1] If export of coin were permitted this bullion would be sent to the Mint, and some, at least, of the coin produced might remain in the country. But neither this nor his further proposal to prohibit the export of bullion was well received, and the shortage of silver coin persisted. During the War of the Spanish Succession, rates of freight and insurance were too high to allow of much traffic in the precious metals. Yet the pull from abroad on English supplies of silver was sufficient to keep its value at a high level. In March, 1711-12, Newton declared that 'gold is overvalued in England in proportion to silver by at least 9d. or 10d. in a guinea, and this excess of value tends to increase the gold coins and diminish the silver coins of this kingdom'.[2] It was, no doubt, with the object of meeting the deficiency of the supply of silver that half a million of the Lottery Loan of 1711 was reserved for those who brought their household plate to the Mint —in spite of the opinion of Newton that the nation's reserves of silver were safer in the hands of individuals, in the form of plate, than in those of the officers of the Mint.[3]

After 1716 the East India Company was offering a higher price for exportable silver than for a corresponding weight of English coin, and there was also a heavy drain of silver to Holland, Denmark, Norway, and Sweden. During the war exports of the metal had varied between a minimum of £193,000 in 1705 and a maximum of £714,000 in the famine year of 1709. In the early years of the peace they were moderate in scale, but in 1717 rose to £1,151,000 and in 1718 to £1,894,000.[4] Hence the country found itself in much the

[1] Mr. Newton's Memorial concerning the proportion of Gold and Silver in Value. Reprinted in W. A. Shaw, *Select Tracts and Documents illustrative of English Monetary History, 1626-1730*, pp. 154-5.
[2] 'Representations of Sir Isaac Newton on the subject of Money.' Reprinted in J. R. McCulloch, *op. cit.*, p. 270.
[3] A. H. John, 'Insurance Investment and the Money Market of the Eighteenth Century', *Economica*, (new ser.) vol. xx, No. 78, p. 138n.
[4] G. N. Clark, *Guide to English Commercial Statistics, 1696-1782*, p. 77.

same condition as before the recoinage, with a silver currency consisting mainly of coins which, by wear and tear, or by clipping and filing, had a value well below that inscribed on them when they came from the Mint.

Complaints, both of shortage of coins of small denomination, and of the deficiency of weight of such of these coins as remained in the currency, were widespread. It was clear that the remedy lay in establishing a ratio between gold and silver closer to that prevailing in Europe (outside the Iberian peninsula) and in the East. This could have been effected by putting less metal into the silver pieces or more into the gold pieces; but a wholesale recoinage would have been expensive and upsetting to trade. It was much simpler to leave the weight of the coins unchanged, and either to raise the denomination of the silver pieces, or to lower that of the gold pieces. For whatever reason, it was the second of these courses that was adopted. On the advice of Sir Isaac Newton,[1] in December, 1717, a proclamation was issued reducing the value of the guinea from 21s. 6d. to 21s. and prohibiting the payment or receipt of gold coins at any higher rate. The effect was to alter the mint ratio from 15·57 : 1 to about 15·21 : 1. This measure, however, proved inadequate: the drain of silver could have been checked only by a reduction that would have brought the ratio close to that in the Far East. As it was, exportable silver bullion still stood at a price higher than silver in the coinage, and a profit was to be made by importing gold, turning it into guineas, and exchanging these for silver coin which could be melted and sent abroad.

If the demand from overseas had been temporary the increased supply of silver bullion resulting from the melting of coin might have brought the market ratio into line with the mint ratio. In fact, it was persistent, and the Mint was powerless. Not only did the undervaluation lead to an export of silver: it discouraged the import, and coining, of the metal. Between 1717 and 1760 little more than £500,000 of silver was offered to the Mint, and during the last forty years of the century only negligible amounts were coined.[2] The public had to make shift with shapeless and debased pieces, and though, as Lord Liverpool pointed out, the consequences of the low *quality* of the coins were less serious than in earlier periods, the

---

[1] 'Representation Third to the . . . Lords Commissioners of His Majesty's Revenue.' Reprinted by J. R. McCulloch, *loc. cit.*, p. 274.
[2] C. Jenkinson, *Treatise on the Coin of the Realm* (1805), pp. 153-4.

deficiency in the *quantity* of silver money was a handicap to the development of retail trade and to the rise of a wage-earning class.[1] As might be expected, the shortage of silver coin resulted in its exchanging for gold at a value higher than that prescribed by law. In 1759 Sir John Barnard declared that bankers generally gave a premium for silver coin in order to be able to supply the needs of their customers. The premium seems to have been high in the months of harvest when silver coin was required for the payment of wages.[2]

Later in the century the gold guineas and half guineas suffered (though in less degree) the same fate as the silver coins. Merchants and manufacturers picked out the heavier pieces and sold them to the bullion dealers and exchange brokers. As Adam Smith put it, 'The operations of the mint were . . . somewhat like the web of Penelope; the work that was done in the day was undone in the night. The mint was employed, not so much in making daily additions to the coin, as in replacing the very best part of it which was daily melted down.'[3] When the foreign balance of payments and rates of exchange were adverse gold poured out of the country. According to law, the English gold coin should have been worth £3 17s. 10½d. an ounce. But in 1763 it was observed that 'the demand for gold in coin is so great that the Jews now give 4 guineas an ounce, so that we may soon expect to have that as scarce as silver'.[4] In these circumstances clipping and sweating was brisk: the guineas that remained in the currency were not only relatively few in number but deficient in weight. In 1773 it was decided to correct the depreciation: some £16½ million of the light pieces were called in and new ones of full weight (129·4 grains to the guinea) were given in exchange. How long they might have preserved their integrity can only be guessed. For the paper issues of the wars with the Americans and the French raised the value of bullion to a height that drew guineas and half-guineas into the melting pot and so overseas.

Worst of all, perhaps, was the state of the copper halfpennies and

---

[1] The evils of debasement of silver had been described by William Lowndes in 1695. They included 'great contentions . . . in fairs, markets, shops and other places . . . to the disturbance of the public peace'; a decline of trade, since before people could make a bargain they had 'first to settle the price or value of the very money they were to receive for their goods'; a rise of the general level of prices; and an unfavourable movement of the foreign exchanges. In the eighteenth century, however, the larger part played by gold was a mitigating influence.
[2] L. S. Sutherland, 'The Accounts of an Eighteenth-Century Merchant', *Economic History Rev.*, vol. III (1932), pp. 380-1n.
[3] *The Wealth of Nations*, vol. II, p. 49.
[4] *Gentleman's Magazine*, May, 1763.

farthings, which, legal tender for sums below sixpence, were supposed to contain an amount of metal equal in value to the figure they bore on the face (less the cost of manufacture) but which, owing to the neglect of the Mint, and the attentions of the filers and clippers, had fallen so low that traders would accept them no longer by tale but only by weight. Between 1702 and 1717, no new copper coins were issued, and (apart from the striking of 200 tons of 'Tower' halfpennies and farthings in 1771-5) the same was true of the forty years after 1754.[1]

In the last decade of the century the skill and ingenuity of Matthew Boulton were called on; and from his mint in Birmingham there poured a stream of twopenny pieces, pennies, halfpennies, and farthings—the first of which were made of exactly two ounces, and the second of one ounce of metal, in order that they might serve the double purpose of currency and weights for the scales of shopkeepers.[2] Soon after the issue of these in 1797, however, the price of copper began to rise, and so the twopenny and penny pieces quickly disappeared into the melting pot. (In 1806 a new and lighter issue was made; but again the price of copper rose, and from the following year to 1821 no legal coins of copper were minted.[3])

The dearth of money of all kinds had important social effects. In a century of economic expansion, manufacturers tended to lock up a high proportion of their resources in buildings, machinery, and stocks of materials, and to hold only a small proportion in ready cash. This proclivity had serious results in times of crisis, for it meant that when bills of exchange were no longer accepted for the payment of debts the manufacturers and merchants had small reserves of hard cash on which to draw. Even, however, if they had wished, employers must have found it difficult to accumulate stocks of ready money. Much time was spent riding about the country in search of cash with which to pay wages, and in the northern and western parts of England the dearth of coin was often acute. In the early days of the century Manchester linen drapers and Rochdale clothiers who provided raw material and yarn to country manufacturers had to make payment in cloth; and at the end of it the Sheffield factors paid the small masters for their cutlery and files in steel and other commodities. It is not surprising that the

---

[1] In 1787, it was estimated, the lawful copper coins in circulation amounted in nominal value to only £322,000. C. Jenkinson, *op. cit.*, p. 211.
[2] *Ibid.*, p. 215.
[3] A. E. Feavearyear, *The Pound Sterling*, pp. 175, 192, 296.

manufacturers treated their own workmen in the same manner, and that the truck system was widespread.[1] Some relief was provided by enterprising industrialists who manufactured their own token coins. The practice (which goes back to the sixteenth century) had been made illegal in 1672. Between 1693 and 1701 private persons were allowed to take blanks to the Mint and to have struck for their own use copper halfpennies and farthings up to a specified maximum amount.[2] In the seventeen-twenties private minting of copper was widespread in Ireland, and in the later decades it grew to considerable proportions in England.[3] In 1786 the Adelphi Cotton Co. countermarked halfpennies and undertook to redeem them at 4s. 6d. each; and in the following year the ironmaster, John Wilkinson, and the copper magnate, Thomas Williams, both put out new tokens of copper. At first these circulated only in the neighbourhood of the ironworks and in Anglesey; but later, when they were made redeemable in London and Liverpool, they were bought by employers and used to pay wages in various parts of the country. The fact that numbers of them were never presented for redemption suggests that they met a real need. It was not long before other industrialists followed suit, and in 1792 it was said that the country was flooded with tokens. Many, both of these and of later issues, were light in weight, and, since they bore no indication of origin, were irredeemable. During the later stages of the Napoleonic War traders' tokens for 1s. and 6d. were in general circulation, and, though attempts were made to suppress them, it was not until after 1821 (when, for the first time, the Mint provided an adequate supply of small change) that the manufacture of private money came to an end.[4]

More important in meeting the needs of the public for cash was the downright counterfeiter, a man who, like Adam Smith's smuggler, deserves less odium than was accorded him by well-to-do people, who thought it no offence to pick out heavy coin for sale to exporters. Counterfeiters abounded in London, and in 1744 complaint was made of the existence of illegal mints in Birmingham. Naturally their activities were concentrated on making forgeries of

[1] A. P. Wadsworth and J. de L. Mann, *op. cit.*, pp. 80-88; T. S. Ashton, *An Eighteenth Century Industrialist*, pp. 38-9.
[2] G. Findley Shirras and J. H. Craig: 'Sir Isaac Newton and the Currency', *Economic Journal*, vol. LV, p. 236.
[3] R. Ruding, *Annals of the Coinage* (1817, Third ed., 1840), vol. II, p. 73.
[4] A. E. Feavearyear, *op. cit.*, p. 296.

the more valuable coins, and it was with those found guilty of uttering pieces purporting to be legal gold coins that the state dealt most harshly: in 1733 when the old worn gold 'broad pieces' were being bought up for recoinage it was made high treason to manufacture copies of them.[1] But the striking of coins of lower value was not neglected, for not only were the penalties on those found guilty less severe, but the shapeless, worn, and defaced silver pieces were easy to counterfeit and there was little risk of detection.[2] During the early years of the century there had been many complaints of the 'raps' that were in circulation in Ireland; and in 1753 it was said that nearly half the copper in use in England was of false coin.[3] An attempt was made to suppress the counterfeiting of copper in 1771, when what had previously been a misdemeanour was made a felony.[4] But the practice persisted, and even tokens such as those issued by Wilkinson in the seventeen-eighties and by the Bank of England in 1811, were counterfeited.

Little is known of the forgers who made a success of their calling, but the *Newgate Calendar* has much to say of the last hours of those who were unlucky. A leading part seems to have been taken by the Irish, who operated in England as well as in their own country. The trade was seasonal and, like many other occupations, was largely suspended in the period of harvest and hop-picking. Year-to-year variations in its activity reflect changes in the needs of the public for coin but, even more, changes in the needs of the forgers. Figures of the prosecutions for counterfeiting (which are available from 1786) show increases in each of the periods of depression, and decreases in years when other opportunities of employment were plentiful.[5]

The eighteenth century was not marked, as many centuries have been, by official debasement of the coinage. Whether or not debasement is harmful must depend on the circumstances in which it occurs. The object was generally to increase the income of the Sovereign; and, when debasement took place in conditions of depressed industry and trade, the expenditure of this increased income drew more people into employment, with little or no effect on prices. When, on the other hand, it occurred in conditions approaching full employment, its effect must have been to raise prices and so

[1] D. Macpherson, *op. cit.*, vol. III, p. 193.   [2] C. Jenkinson, *op. cit.*, p. 183.
[3] R. Ruding, *op. cit.*, p. 80.
[4] D. Macpherson, *op. cit.*, vol. III, p. 512.   [5] Report of the Royal Mint (1849).

alter the distribution of wealth without doing anything to real income.

Precisely the same thing was true of the actions of the clippers, counterfeiters, bankers, and issuers of tokens. They, too, sought to increase their incomes: the expenditure of these incomes created employment and incomes for others. Whether their actions were anti-social in their effects depends on the conditions in which they were taken. If it is true that the counterfeiter came into being because of a shortage of coins, some at least of the blame for counterfeiting must rest on the monetary authorities. The action of the counterfeiters in melting down the better copper coins issued by the Mint to furnish material for their own lighter coins, was reprehensible, and may, it has been suggested,[1] have been one reason why the Mint ceased to issue copper coin. But, this apart, there is little evidence that either the clippers or counterfeiters did much harm, and some that they did good, at times when England was short of hard money and labour was underemployed.

The remedies for the varied ills to which the monetary system was subject were discovered only by much trial and error. A first step was taken in 1717 when a ratio favourable to the circulation of gold was established. In this year Sir Isaac Newton expressed (though he did not publish) the opinion that 'gold is now become our standard money, and silver is a commodity which rises and falls here in its price as it does in Spain'.[2] But in the minds of most people the silver pound was still the measure of value. 'The laws, the language of the country, the common consent, and common sense of all men have unanimously concurred in making silver our only standard', wrote the Assay-Master, Joseph Harris, in 1752.[3] Six years later Adam Anderson[4] declared that 'silver ever has, and probably ever will, hold the prerogative of being the fixed standard, gold being always valued by silver, but not silver by gold'. Again, in 1776, Adam Smith observed that 'In England . . . and . . . in all other modern nations of Europe, all accounts are kept, and the value of all goods and of all estates is generally computed in silver: and when we mean to express the amount of a person's fortune, we seldom mention the number of guineas, but the number of pounds sterling which we suppose would be given for it.' Nevertheless, Smith recognised that, once the state had decreed that the guinea should exchange for so many shillings, the difference between the

---

[1] G. Findley Shirras and J. G. Craig, *loc. cit.*, p. 238.　　[2] *Ibid.*, p. 234.
[3] J. R. McCulloch, *op. cit.*, p. 488.　　[4] D. Macpherson, *op. cit.*, vol. III, p. 56.

metal that was, and that which was not, used as the standard became 'little more than a nominal distinction'.[1] In fact, the sovereignty of silver was being steadily diminished. During the recoinage at the end of the seventeenth century, when silver coin was for a time out of circulation, men had become accustomed to the feel of the guinea. In 1717 gold was given the status of legal tender. Merchants had for long found it convenient to use gold in their larger transactions and to offer silver only for small payments; and, as other classes of people grew richer, they followed suit. By 1760 the crown had almost disappeared, the number of half-crowns was small, and the silver in circulation consisted mainly of worn shillings and sixpences. There is a general tendency for the chief medium of exchange to become the standard of value; the heavy gold piece seemed more suitable as a measuring rod for internal transactions, and, as time went on, it took the place of silver in the settlement of balances with other nations.[2] It was not by statute but by what Lord Liverpool called 'the disposition of the people' that the gold standard was silently established in England.[3] The change in the status of the two metals was openly acknowledged in 1774 when the legal tender of silver, by tale, was restricted to payments not exceeding £25.[4] It was made final in 1816 when (mainly at the instance of the second Lord Liverpool) gold was declared to be the sole standard and full legal tender, and a new coin, known as the sovereign (containing 123·27 grains of gold, eleven-twelfths fine) was put into circulation.

II

Supplies of gold and silver coin were supplemented by currency in other forms. Shortly before the eighteenth century opened the creation of the national debt had brought into being a mass of securities bearing more or less fixed rates of interest. Some of these, including exchequer bills, navy bills, and lottery tickets (as also the short-term obligations of the East India Company, the Bank of England, and the South Sea Company) could be used to settle accounts between individuals, and may perhaps, therefore, be thought

[1] *The Wealth of Nations*, vol. I, p. 34.
[2] C. Jenkinson, *op. cit.*, p. 186.
[3] A. E. Feavearyear, *op. cit.*, pp. 142-6.
[4] C. Jenkinson, *op. cit.*, p. 142. The provision was allowed to lapse in 1783 but was re-enacted in 1798.

of as falling within the somewhat shadowy boundaries of 'money'. Even the long-dated securities had some effect on the volume of purchasing power. For the fact that any holder could dispose of them through stock dealers and so obtain cash (though at the expense of the cash holdings of others) meant that men were less concerned than their fathers had been to keep quantities of coin, bullion, and plate locked up in safes or buried in their orchards and gardens.

More important was the creation of institutions which, whether they took the name or not, exercised the functions of banks. When, in 1718, Richard Ford and Thomas Goldney became the principal partners in the Coalbrookdale ironworks, it was decided that Goldney, who lived in Bristol, should collect the debts due to the firm and hold the money in a fund which, in the correspondence between the partners, was referred to as 'the bank'.[1] Other examples could be given of the use of the term to denote a stock of money: they call attention to the prime essential to the setting up of banks in the more generally accepted sense. Bankers, however, are not simply dealers in coin. They are men whose debts (or promises to pay) are widely accepted by others as though they were legal money: it is part of their business to create means of payment other than hard cash. They must be (or be thought to be) men of substance and integrity; and some of the bankers of this period owed little to anything but their own resources and character. But, as at other times and places, some special connection with Government, or a position in the economic system that gave control over the money of others, was often the occasion of the setting up of a bank.

When under the Tonnage Act of 1694, a company known as the Bank of England received a charter of incorporation, it undertook to lend to the Government the sum of £1,200,000. In return, it was given the right not only of receiving money on deposit from the public and lending this at interest, but also of making loans in paper, which it brought into being at will. Of the original loan to the Government only £720,000 was in cash: the remainder consisted of what were known as 'sealed bills'. Throughout the century the bulk of the loans and discounts made to the public took the form of written (and, later, printed) promises to pay, which were spoken of as cash notes, or simply as notes; and in this way the Bank of England made substantial additions to the media of exchange, at some benefit to its proprietors. This was not its only function: it supplied

[1] A. Raistrick, *op. cit.*, p. 8.

gold and silver to the Royal Mint, made remittances overseas for the Government, received subscriptions to state loans, sold exchequer bills, and looked after the public accounts. It lent to the other chartered companies and to a select number of merchants and traders who lived in, or had close connections with, London. It had, however, little contact with the provinces, and few of its notes found their way to the industrial areas of the east, the south-west, and the north. As Sir John Clapham remarked, it lived up to its nickname of the Bank of London.[1]

Most of the services rendered by the Bank were a continuation and expansion of those provided, from earlier times, by individual merchants, brokers, scriveners, and goldsmiths of the City.[2] In the seventeenth century the goldsmiths, who were well-to-do and possessed safes and strong rooms, had developed into bankers in the modern sense of the word. The receipts they gave for moneys lodged with them passed from hand to hand subject only to endorsement, and became, in effect, bank notes. Anyone who had a deposit with them might write a letter, or a draft, giving instructions for the transfer of a sum of money to the account of some other person; and such letters or drafts, though also described as notes, were in effect cheques.

In the later decades of the seventeenth century, the older goldsmith houses (those of Thomas Viner, Edward Backwell, Francis Meynell and others) encountered competition from a new set of private banks, established by men such as Francis Child, Richard Hoare, and Charles Duncombe. In the period 1750-65 the London banks numbered twenty or thirty, in 1770 fifty, and in 1800 seventy.[3] As Mr. Joslin points out,[4] there was some differentiation of function. Some banks, like those of the Hoares, Coutts, Childs, and Drummonds, specialised in business with the aristocracy, the gentry, and well-to-do lawyers (to all of whom they made loans on mortgages or bonds) and in dealings in government securities. Other banks (such as Martins, Curries, Glyn Mills, Mastermans), which arose mainly in the second half of the eighteenth century, concentrated rather on making call loans to members of the Stock Exchange, discounting bills for traders and manufacturers, and acting as correspondents for the country banks that were springing up at

[1] Sir John Clapham, *The Bank of England*, vol. I, p. 215.
[2] R. D. Richards, 'The Pioneers of Banking in England', *Economic History*, No. 4 (1929), p. 485.
[3] Sir John Clapham, *op. cit.*, vol. I, p. 158. [4] *Ec. Hist. Rev.*, vol. VII, (sec. ser.) no. 2.

this time. Many of these London houses issued their own notes, and all of them, through the facilities they offered for discounting bills, played a part in increasing the supply of purchasing power in the metropolis, and, to a less extent, in the country.

Outside the London area formal banking was a relatively late growth, though many of the functions of the banker had for long been exercised by men who were content to describe themselves simply as merchants or traders. The country dealer or manufacturer often needed more capital than could be obtained from his own resources or those of his partners and friends. He required facilities for making payment to creditors in other parts of England. And he needed supplies of currency for the purchase of materials and the payment of wages. As has already been pointed out, much of the capital of the trader and industrialist was provided by merchants who had acquired wealth in domestic or overseas trade, and such men were also well-equipped to provide means of remittance. Many illustrations are offered in the illuminating study of Mr. Wadsworth and Miss Mann. In the late seventeenth century, Thomas Marsden of Bolton set up a London house through which he bought his raw materials and disposed of his yarn and fustians. Having a stock of money in London, he was able to supply bills of exchange to traders who had payments to make there. And since a bill on London found ready acceptance in other places, he could equally supply means of remittance to all parts of the country. (For this service he charged what, in the circumstances of the time, must be regarded as the low commission of 5s. for a bill of £100 payable in one month's time.) Having a stock of money in Bolton, he was also in a position to buy, or discount, bills which his neighbours had received in payment for goods sent to customers in other places, and so to provide money for local use.

In this case, as in many others, banking was merely an extension of the functions of the trader. Many houses, both in London and the provinces, engaged in it without shedding their other functions. In 1757, J. and N. Philips, manufacturers of Tean in Staffordshire, wrote to Thomas and Mitchell, a firm of London dry-salters, saying, 'If it be agreeable we shall send you all our bills and draw on you at our Conveniency, allowing you a Commission for your Trouble.'[1]

Retailers and innkeepers, no less than traders, were often called on to exercise the functions of the banker. They were well known

---

[1] A. P. Wadsworth and J. de L. Mann, *op. cit.*, p. 298.

to their fellow-townsmen, and could act as intermediaries between those who wished to borrow and those with funds to invest. William Stout of Lancaster recorded that the Quaker shopkeeper, Henry Coward, to whom he was apprenticed in 1680, was held in such repute that 'any who had money lodged it with him to put out to interest or make use of'. From the nature of their calling retailers acquired stocks of coin: they were able to supply ready money to farmers and manufacturers who needed it to pay wages; receiving credit from wholesalers, they were able to extend it to customers; and they were used to handling bills and promissory notes. How a shopkeeper might profit from the situation he occupied may be illustrated by the career of Peter Davenport Finney, who, about 1755, set up as a confectioner in Manchester and within a few years became the owner of a grocery concern with both retail and wholesale connections. According to his brother,[1] one reason for Finney's success was 'his custom of buying with ready money, which the great returns of his retail business supplied him with'. This did not mean, however, that he made his payments in hard cash, 'for *that* he first changed into bills, for which he had a discount, and afterwards[2] made his payments with those bills as ready money, for which he also had a discount for ready money'. Finney had taken the first step towards banking: he had become a discounter of bills. If he did not go further along the path it was only because he was already well enough off to buy from his brother the reversion of the family estate and to settle down as a landed gentleman at Wilmslow. There can be little doubt, however, that it was a similar process that led his neighbour in Manchester, John Jones, to add to his business as a tea dealer that of a banker, and so to bring into being the house of Jones Loyd with which, in the following century, the name of Lord Overstone was closely associated.

Among others who were able to put to profitable use money that passed through their hands were the officials of turnpike trusts and canal companies, and the local receivers of taxes.[3] Thomas Marsden, referred to above, must have owed much of his success to the fact that he was the Returner of Revenue for the county of Lancaster. The growth of taxation during the wars of William and Mary called for a considerable remittance of money to London; and, since the

[1] Samuel Finney, *An Historical Survey of the Parish of Wilmslow*.
[2] I.e. when the bills had matured.
[3] L. S. Pressnell, 'Public Monies and the Development of English Banking.' *Ec. Hist. Rev.*, vol. V (second ser.), pp. 378-97.

local collectors of the Land and Assessed Taxes and Stamp Duties (unlike those of the Excise) were allowed to retain, for as long as a year, the money they drew from the public, they were exceptionally well placed to act as bankers. There was nothing irregular in this. High officials of His Majesty's Treasury drew profit from the use of public money, and the privileges of the tax-collectors are to be thought of as part of the remuneration of an otherwise poorly paid body of piece-workers. The receivers of revenue in various parts of the country discounted bills and lent on mortgage to local agriculturists and manufacturers. When they had to remit to London they bought bills of local traders, or, if the price of these was high, made their remittance in coin. (There were specie points in domestic, as well as in overseas trade.)

A list of country bankers drawn up in 1784 contains the names of at least half-a-dozen receivers of taxes; and, during the period of heavy government spending that began in 1793, the number of those who combined the two functions increased rapidly. It should be added, however, that not all of these had begun their careers as revenue officers. For if the collecting of taxes offered facilities for banking, the local bank (in which public moneys were often deposited) might also be a school for tax collecting. All that the lists show is that the two occupations were frequently carried on in conjunction.

Yet other recruits to banking were found among the country attorneys who received money on trust and were knowledgeable as to investment: when, about 1782, Thomas Lyon and Joseph Parr set up the Warrington Bank they took as their partner a local solicitor, Walter Kerfoot.[1] Closely associated with the lawyers were men known as money scriveners whose business it was to act as intermediaries between those with funds to invest and landlords and others who sought to borrow on mortgage or personal bonds. It was not a long step from acting in this way as an agent to banking on one's own account.

Finally, in the later decades of the century, the larger manufacturers of cotton, iron, copper, and other commodities began to take an interest in the formation of banks. Such men were constantly in need of capital for the extension of their works and of cash for the payment of wages. Some were able to get help from merchants and bankers in the metropolis: for example, Matthew Boulton from

[1] T. E. Gregory, *The Westminster Bank*, vol. II, p. 24.

Vere, Williams and Vere, Samuel Oldknow from S. and W. Salte, and John Wilkinson from Smith, Wright and Gray.[1] But it was obviously more convenient to conduct day-to-day transactions with banks nearer to their own establishments. Hence, in 1778, Boulton (who had extensive interests in Cornwall) opened an account with Elliott and Praed of Truro; in 1786 Oldknow began to make use of the services of Evans and Sons of Derby (who also financed Arkwright); and, in 1793, Wilkinson not only opened an account, but became a partner, in the Shrewsbury bank of Eyton and Reynolds. Many industrialists were active in promoting country banks. In 1765 the ironmaster, Samson Lloyd, joined with a button maker, John Taylor, to establish a bank in Birmingham; in 1790, or earlier, the calico printer, Robert Peel, instituted the Manchester bank of Peel, Greaves & Co.; and in 1792 the Walkers of Masboro' took a leading part in setting up the bank of Walker, Eyre and Stanley of Rotherham and Sheffield.[2]

The country banks (though not country banking operations) were a product mainly of the second half of the century: only about a dozen of them were in existence before 1750. A list compiled in 1797 records the names, situation, and London correspondents of 334 country banks in England and Wales; and thirteen years later the number had doubled. No annual figures of new formations are available, but most banks seem to have been set up at times of rapidly expanding trade, such as 1750-3, 1762, 1765-6, 1770-3 and 1789-92. There was, it is true, a spate of new banks in 1774, when industry was sorely depressed.[3] But this was a year in which guineas and half-guineas were being called in for recoinage, and when there was, therefore, a special demand for the services of banks, both to collect the coins and provide alternative currency. If the creation of banks was, at once, a symptom and a cause of inflation the subsequent disappearance of many was equally both a symptom and a cause of deflation. One reason for the weakness of the country banks is that the Bank of England had been given a monopoly of joint-stock banking south of the Tweed. The provincial banks, like those of London, were generally partnerships of two or three men, and their

[1] H. W. Dickinson and Rhys Jenkins, *op. cit.*, p. 50; G. Unwin and others, *op. cit.*, pp. 177-8.
[2] L. H. Grindon, *Manchester Banks and Bankers*, p. 64; R. E. Leader, *The Sheffield Banking Company Limited*, p. 6.
[3] Among the banks that came into existence at this time were the well-known houses of the Gurneys of Norwich, the Backhouses of Darlington, and Parker and Shore of Sheffield.

stability might be endangered by the death, or loss of reputation, of any one of these. Their fortunes were bound up with those of a narrow range of local industries. In 1782 the Gurneys of Norwich opened branches at Yarmouth, Halesworth, Lynn, and Wisbech, but such geographical extension was unusual: the great majority of private banks had only a single office. Little information exists as to the distribution of assets, but it is known that many banks held unduly small cash reserves against their note issues and deposits. They relied on their correspondent banks in London to provide them with coin, and if this were not forthcoming they were forced to close their doors. Some bankers used an undue proportion of the sums deposited with them to finance manufacturing or trading concerns of their own: though assets might more than cover their liabilities they could not always be easily realised, and, in these circumstances, an unexpected demand for cash might spell disaster and disgrace. It would be wrong, however, to assume that all country banks exhibited these weaknesses. Most of them rode safely through severe storms and carried their clients with them. By making loans on mortgage or bond they helped to build up many large industrial concerns, and by issuing notes, and supplying bills and drafts, they played an important part in alleviating local shortages of currency.

The country houses fall into two broad classes. In the rural areas their chief function was to act as banks of deposit. Farmers and traders brought in bills they had received for their products: sometimes they would discount these for coin or notes, but often they were content to build up a balance. When the bills matured they were sent to the correspondent for collection, and the rural banks piled up funds in London. They were usually willing to pay a commission to the London bank for its services in finding profitable investments for these, and, in the later decades of the century, specialist bill brokers were serving as intermediaries between banks with money to invest and banks or merchants who wished to discount bills. In the growing manufacturing areas, on the other hand, the chief function of the banks was to provide advances to their clients. The demand for means of payment normally exceeded the supply, and the local banks had to draw on their correspondents in London. In order to do this they remitted to the metropolis parcels of bills which had not yet reached maturity. It was the business of the London banks to use the balances of the rural banks in discounting these; and hence to transfer purchasing power from areas in which it was relatively plentiful to those in which it was relatively scarce. Much of the investment in manufacture was provided, in this way, by the savings of agriculturists in other areas.

Another broad distinction is between the banks that lent in their own notes and those that did so only in coin, Bank of England notes, or bills and drafts. Most of the rural banks were note issuers, as were also the banks in the industrial areas about Newcastle, Norwich, Bristol, Sheffield, and Birmingham. But in the most rapidly growing region of manufacture—that of Lancashire and a large part of the West Riding—few of the banks issued their own notes. For, in this compact and closely knit area, the promissory note and the bill of exchange, drawn by one manufacturer or trader on another, had long been found to meet satisfactorily the needs of local commerce and to serve, in large measure, as a substitute for coin.

By means of a bill, purchasing power could be transferred by one man to another, under conditions of repayment plainly set forth and generally understood. Unlike the coin or bank note, the bill could be sent from place to place without danger of theft. It could pass from hand to hand without formality other than endorsement, and each person who put his name to it added to its security. Any holder could get coin or other currency by discounting it: as a security it was highly liquid. Its status had been raised, and defined, by Acts passed about the end of the seventeenth century, and notably by one of 1698 which laid down the procedure for protesting non-payment.[1]

Inland bills and promissory notes played a considerable part in the trade of all parts of England and Wales. But nowhere had their use extended so far as in the north-west. The ubiquity of the bill was probably the reason why in this area formal banking made its appearance relatively late: there seems to have been no bank in Manchester before 1771, or in Liverpool before 1774;[2] and the list of 1797 contains the names of only about a dozen bankers in the county of Lancaster. Apart from granting mortgages or lending against securities, these banks confined their activities to discounting bills for those who needed ready money, and supplying bills, or their own drafts on London, to those who had payments to make in other areas. These facts may help to explain the relative immunity of Lancashire and the West Riding from the panics of those parts

[1] For these see R. B. Westerfield, *op. cit.*, pp. 390-1.
[2] A. P. Wadsworth and J. de L. Mann, *op. cit.*, p. 142.

of England in which the paper currency rested on the reputation not of the traders, but of a number of small, and not always well-managed, banks of issue.

It is not to be thought, however, that the use of bills and promissory notes gave exemption from currency difficulties. Not all bills arose out of commercial transactions: a man might arrange to draw on another simply as a means of raising funds to meet his personal expenses, or for speculative purposes. There is nothing essentially dishonest or objectionable in accommodation bills. But in times when the outlook was bright there was a tendency for their volume to rise unduly, and they played a special part in inflationary movements.

A feature of the circulation in the north-west was the large volume of small bills and promissory notes created by manufacturers for the payment of labour. Some were for amounts as low as eighteen pence or a shilling, and it is said that bills for sixpence were not unknown. The wage-earners who received them passed them to local retailers who, when they had accumulated a sufficient amount, returned them to the manufacturers in exchange for larger bills drawn on London. The small bills and notes served a useful function when, as in 1773, coin of the realm was hardly obtainable. But it is rarely safe to entrust an employer with the creation of the currency with which he pays his debts to his workers: there was a strong temptation to exercise the power even when supplies of legal money were available.

Some of the small bills and notes bore on the face words that limited the claims of the holder against the issuer; and the poor, who understood little of such matters, often found they could turn them into cash or commodities only at a discount. In 1775, at the instance of Sir George Saville (who represented the county of York), an Act was passed to restrain the negotiation of promissory notes and inland bills for sums less than twenty shillings;[1] and two years later the prohibition was extended to notes and bills for amounts under £5.[2] It is by no means certain, however, that these measures were of benefit to the workers. For it was still possible for the employer to pay wages in paper of larger denomination. He might give a bill for £5 or more to a group, leaving them to divide it among themselves at the public house. Or he might lengthen the interval between pay days, until the wage due to the individual reached, or

[1] 15 Geo. III, c. 51.   [2] 17 Geo. III, c. 30.

exceeded, five pounds. In the 'eighties Samuel Oldknow managed to find coin to pay the poorer of his spinners fortnightly, but the more well-to-do had to wait for a month or two months before receiving their wages in the form of bills.[1] (His employees at Anderton had to pay 3*d*. in the £ to the local shopkeepers for turning the bills into cash.[2]) Whether legally or not, in the crisis of 1793 he paid his factory workers in notes or orders on the shop he had set up at Mellor, and these seem to have passed from hand to hand in the locality.[3] A few months earlier—without the excuse of a crisis—John Wilkinson had acquired supplies of the depreciated French assignats, and had paid these, countersigned by his clerk, to his workers at Bersham, near Wrexham. He was the brother-in-law of the radical, Unitarian, Dr. Priestley; many of his workers were Methodists or free-thinkers; and lectures on Paine's *Rights of Man* were being delivered in Wrexham. These irrelevant facts were adduced by a Welsh squire as evidence of the sinister nature of Wilkinson's experiment with the currency; and an act was hurriedly passed to suppress the circulation of all French notes in England. When, however, in the financial crisis of 1797, the ban on the issue of small bills and notes was lifted, Wilkinson immediately printed cards for amounts as small as a shilling, sixpence, and threepence and paid these out as wages.[4]

In spite of the fact that, from this time, notes of small denomination could be obtained from the Bank of England, as well as from most of the country banks, several large manufacturers followed Wilkinson's example; and throughout the war (and, indeed, for a quarter of a century after it ended) the small bill continued to play a large part in the circulation of industrial Lancashire.[5]

Enough has been said to indicate the wide variety of the agencies for the issue of paper money. At first glance it might appear that the total volume of paper was determined solely by the cupidity or caprice of irresponsible issuers. Large creations of notes and bills helped to create booms, and sudden restrictions led to crises and

[1] G. Unwin, *op. cit.*, p. 70.   [2] *Ibid.*, p. 50.   [3] *Ibid.*, pp. 179-83.
[4] In March, 1797, he wrote Boulton and Watt as follows: '. . . am engaged in preparing small notes for my workmen as change, similar to what I issued in '73 and '74 previous to Sir George Savile's Act. That was a measure I adopted on the then great scarcity of silver, which since has been plentifully supplied by the carriers of bad money. *Good* notes will cure the evil of base metal better and more effectually than the gallows.' I am indebted for this quotation, as well as for details in the text, to Dr. W. H. Chaloner.
[5] T. S. Ashton, 'The Bill of Exchange and Private Banks in Lancashire, 1790-1830', *Ec. Hist. Rev.*, vol. XV, Nos. 1 and 2 (1945), pp. 25-35.

depressions. There can be little doubt that with a more orderly system of issue England might have avoided much discomfort. But, at least, the currency never fell into chaos, as it did in France. There was a system, even if its members were loosely articulated. The supply of bills was not unaffected by the supply of notes in which payment might ultimately have to be made. Though the country bankers may not have had clear-cut ideas as to the proper relation between liabilities and reserves, their issue of notes and drafts was not made without consideration of their balances in London. And though, until the end of the century, few London banks held accounts at the Bank of England, their policy was not determined without reference to the availability of Bank notes in the metropolis. Ramshackle as the structure might be, it served the needs of the day more adequately than some of its historians would lead one to believe.

III

If some merchants evolved into bankers others became dealers in foreign exchange. A man who had a claim to money abroad could draw a bill of exchange on his foreign debtor and sell this for sterling to another man who wished to make payment abroad. Both drawers and remitters (as the two were respectively called) could, however, save themselves trouble and expense by getting into touch with an exchange broker who dealt in bills for a commission. An exchange broker was a man who (often as a result of trade in commodities) had acquired stocks of money or credits in foreign parts, as well as in London, and was thus in a position to transfer funds, in either direction, for others. By the end of the seventeenth century the market was well organised and highly competitive. And from 1697 rates of exchange on sixteen European centres were printed twice weekly and offered to the public by J. Castaing, a free broker, at Jonathan's coffee house in London.[1]

Obviously the quality of the moneys of the countries concerned had a bearing on rates of exchange. 'Trading nations', wrote Thomas Prior[2] in 1729, 'assay their coins, and by their assays and weights know how many pieces of the money of one country are equal in

---

[1] A copy of these may be consulted at the Library of the Stock Exchange. For the years 1718-36, and 1747-1811, monthly rates (based on Castaing's figures for the early years, and on Lloyd's List for the later years) are given in *B.P.P.* (1810-11), vol. X, Paper 43.

[2] J. R. McCulloch, *op. cit.*, p. 315.

value to any certain number of pieces of another, and by that means they fix the par of exchange among themselves.' In calculating the par with Amsterdam, Hamburg, Antwerp, and most other centres of northern Europe, it was the practice to treat sterling as the invariable unit and to express its value in terms of foreign money. On the other hand, in estimating the par with Paris, Madrid, Lisbon, Genoa, Leghorn, and other centres of southern Europe, it was the custom to take the foreign coin as the invariable unit and to express its value in terms of shillings and pence sterling. The rate on Dublin and on colonial centres that made use of coins similar to those of England took a different form. The currency of Ireland consisted of guineas, shillings, and pence, but each of these was officially worth only 12/13ths that of the corresponding English piece: English money was said to be at a premium of 1/12th or $8\frac{1}{3}$ per cent., when the exchange on Dublin was at par.

A par of exchange arrived at by comparing the weight of metal in the full-weight coins of, say, England and France, was of little practical significance. For eighteenth-century mints were so inaccurate in workmanship that even new coins of a particular denomination varied one from another in weight; and after they had been in circulation for a few years the divergencies became greater. What mattered to the dealer was the weight of silver or gold he could get from the foreigner in exchange for a fixed sum of the money of his own country. It was for this reason that at Amsterdam, Hamburg, Venice, and some other centres, payments could be obtained either in coin or, if desired, in bank money, the metallic value of which was constant. There were, therefore, two quotations of rates between London and these centres, one denominated 'money' and the other 'banco'.[1] In most countries, however, payment was always made in common money, and hence rates of exchange were intimately affected by changes in the weight of metal in the coins in circulation, or in the weight of bullion they would purchase.

If an exchange broker found that the demand for bills on a foreign centre was exceeding the supply so that his funds (or credits) in that centre were diminishing, he offered less of the foreign money for a pound sterling: if, on the other hand, supply exceeded demand, he offered more. There were, however, limits to the upward or

---

[1] The percentage difference of bank over current money was known as the *agio*. It varied with the weight and fineness of the money in circulation. In time of war when ready money was needed to pay the troops it tended to fall. *Ibid.*, p. 20.

downward movement of the rate. For it was usually possible to make payments between two centres in the form of bills on a third centre. In particular, bills on Amsterdam had an international currency. The fact that no weekly rates were quoted in London on Stockholm, Danzig, Riga, and St. Petersburg, did not mean (as some have supposed) that no bills were used in this trade. For the Dutch had extensive dealings with the Baltic and quoted rates on all the leading ports. An Englishman who wished to make, or receive, payment for transactions with this area could do so by means of bills on Amsterdam (though there is evidence that bills drawn on London were often employed). Similarly, an English broker who found the quoted London rate on any European centre unfavourable might use Amsterdam bills (or for that matter bills of any other place) if by so doing he could make a profit or escape a loss. Hence it came about that the London rates on all foreign commercial centres moved closely together.

At times when foreign rates were unfavourable to Britain it was possible for the English trader or investor to make payment by shipping gold or silver, either directly or through bullion dealers (who were generally spoken of simply as 'the Jews').[1] To do so, however, involved expenses of freight, insurance, and commission, and the rate of exchange at which it paid to remit in this way, instead of paying the price demanded for a bill, varied with the kind of metal available. What were later known as the specie points were not fixed. A moderate fall in the value of sterling was sufficient to lead to a remittance in gold bullion. If none of this could be obtained, sterling might fall a little further until it paid to incur the cost of shipping silver bullion. If the supplies of this on the market were exhausted, another fall would make it profitable to export foreign coin from England. And if no foreign coin could be found, yet a further unfavourable movement of rates of exchange might take place until it became worth while to bear the cost of sending English coin abroad—a cost that included the risk of penalties for an infringement of law.[2] Whether remittance was by bill or by metal was not a question of geography, but of the prevailing rate of exchange. Since, in the last resort, the precious metals in one form or other

[1] In 1742 it was said that if the rate fell below ten guilders sixteen stivers 'the remitters of the Jews will certainly send over our foreign gold and silver, or melt down and send over our coined gold and silver in order to bring the exchange up again to that standard'. *Parliamentary History of England*, vol. XIII, p. 18.
[2] C. Jenkinson, *op. cit.*, p. 179.

could be shipped out, the movements of rates were far smaller than those to which inconvertible currencies have given rise in our own day.

The chief disturbances of the exchanges were the result of political hostilities. In time of war the cost of transporting and insuring gold and silver was high, and hence it was possible for rates to rise and fall to a greater extent than in time of peace. From the beginning of the War of the Spanish Succession sterling fell in terms of most European currencies, but the extent of the depreciation was small. In France budget deficits were such as to lead to a substantial reduction in the value of the currency; and when peace came in 1713 English traders were faced with acute competition by reason of a devalued French crown. The position was clearly stated by Sir Theodore Janssen in 1713:

'Before the War, if I bought any commodity in *France*, which cost me a livre, I paid eighteen pence, *English* for it, as is well known to every body that had dealings there; if I buy now the same thing in *France* for a livre, I pay but one shilling for it, by which means all their manufactures are rendered so very cheap to us, that if there was but moderate duties upon their importation, we should immediately be overwhelmed with *French* commodities. For as their work-men receive no more sols or pence for their days-work or wages than they did formerly, they sell their cloth, paper, and linnen for no more sols than they used to do. . . .

'On the contrary, when the *French* bought any thing of us before the war, if it cost them one pound sterling, they paid but thirteen livres for it; and if they buy now the same thing for one pound sterling, they pay twenty livres: which renders every commodity we have so very dear to them, that 'tis hardly possible they should take any thing from us, but what they have an absolute necessity for.'

Janssen, it may be said in parenthesis, was a strong opponent of trade with the French. 'France produces nothing that is necessary or even convenient, or but which we had better be without', he wrote, adding (as though a serious thought had just struck him) 'except claret.'[1]

No such devaluation occurred in England. But as coins lost weight, by wear and tear or clipping, their purchasing power in bullion, and hence in foreign money, tended to fall. The effect was less marked,

---

[1] T. Janssen, *General Maxims in Trade, particularly applied to the Commerce between Great Britain and France* (1713), p. 15.

192   *An Economic History of England: the 18th Century*

however, than it would have been if the English coinage had consisted solely of silver. According to Adam Smith, 'the gold, that part of it at least which circulated in London and its neighbourhood, was in general less degraded below its standard weight than the greater part of the silver'.[1] So long as the guinea maintained its value rates of exchange were hardly affected by the lightness of the silver money. But by the late 'sixties the guinea itself was in poor shape. Hence the market price of standard gold rose, and the general deficiency of the gold coins in circulation was so notorious, 'that it was estimated in all our exchanges with foreign countries'.[2]

Apart from such influences, rates of exchange varied with the balance of payments. As has been pointed out, the official statistics of imports and exports do not represent current values, and it is impossible, therefore, to say anything about the balance of commodity trade in particular years. It is highly probable, however, that at periods of deficient harvests, when exports of grain ceased and imports rose, the balance of commodity trade was against this country. In all but one of the nine periods of dearth[3] the exchanges fell to low levels: the single exception, the harvest year 1756-7, saw the outbreak of a major war and (for reasons given below) the forces favourable to sterling arising from this were stronger than the adverse forces proceeding from the market for grain. Nor was it only the exchanges of nearby countries, such as France, the Netherlands and Ireland, that moved adversely. Writing from America in 1740, a merchant, Robert Pringle,[4] observed that 'Famine in Europe created an enormous demand for South Carolina rice to such an extent that in Charleston the Exchange fell from £800 per £100 (sterling) to £700'.

If the adverse balance were pronounced the fall in the rates of exchange was such as to result in a noticeable rise in the price of imported goods, and this, as Jenkinson pointed out, gradually extended itself to other commodities and finally reached 'even the most common necessaries of life'.[5] But the movement was generally

---

[1] *The Wealth of Nations*, vol. I, p. 36.

[2] C. Jenkinson, *op. cit.*, pp. 3-4.

[3] 1709-10, 1727-8, 1739-40, 1756-7, 1767-8, 1771-2, 1782-3, 1795-6, and 1799-1801.

[4] I am indebted to Lady Haden Guest for this quotation from her (unpublished) work on *Robert Pringle and the Charleston Trade, 1737-45*.

[5] C. Jenkinson, *op. cit.*, pp. 3-4.

self-corrective. It led to an export of bullion[1] and coin, and this, no doubt with a restriction of credit, usually brought prices down and so discouraged further imports and encouraged exports.[2]

Of greater importance than the balance of commodity trade was the movement of funds into, and out of, Britain. It may be that the part played by Dutch investors in the finance of British government and trade has been exaggerated, but there can be no doubt that it was considerable. When the yield on safe investment in London was one or more per cent. above that in Amsterdam, capital flowed into Britain and the rate of exchange moved in a favourable direction: when the difference narrowed or disappeared, capital flowed back and the rate of exchange fell. There is reason to believe that investment by the Dutch helped to maintain the value of sterling for a short period after the end of the wars of Anne in 1713. But when interest rates in England fell between 1715 and 1717 much of the Dutch money was repatriated and the exchanges moved against Britain. It may be that a similar movement of Dutch capital is the explanation of the relatively low value of sterling in the late 'twenties and early 'thirties. Certain it is that the issue of a new loan at attractive rates drew in Dutch money and raised the value of sterling in 1759; and withdrawal of holdings led to a decline of rates in 1777, 1783, and 1791-2. Investment by Englishmen overseas was probably of less significance. But it is possible that the export of capital to India may explain the fall of the exchanges in 1768-71, and the cessation and reversal of this movement may account in part for the recovery in 1773. Interest payments to foreigners put no strain on the exchanges so long as the flow of capital was towards England; but when this was reversed, the necessity of remitting dividends, to an amount that was estimated at £500,000 or £600,000 a year in the middle of the century, may well have tended to depress the sterling rate.[3]

On the other hand, it would be wrong to think of sterling as being dragged at the heels of Dutch finance. Britain had a hard currency, and large investments overseas. Apart from the commodity trade, she had a substantial claim on foreigners for the services rendered

[1] In 1702 Isaac Newton wrote: 'The safety and increase of the coyn depends principally on the ballance of trade.' He continued: 'If the ballance of trade be against us the money will be melted down and exported to pay debts abroad and carry on trade in spight of laws to the contrary, and if the ballance of trade be for us such laws are needles and even hurtful to trade.' W. A. Shaw, *op. cit.*, p. 156.
[2] The process is clearly set forth in the Bullion Report of 1810.
[3] Evidence of Mr. Bunce, *Parliamentary History of England*, vol. XIII, p. 39.

by her shipowners and underwriters. It might have been expected that any major political crisis affecting her would have led to an unfavourable turn of the exchange; but in fact the opposite was usually the case. In 1701 when the French seized Antwerp, and it seemed probable that England would be involved in war, the rate on Amsterdam soared from 34-7 to 37-3 within a few weeks; and, though there was a subsequent fall, the exchange remained relatively high until, in May, 1702, hostilities actually broke out. (Thereafter, it fell, reaching a low point of 33-10 in August, 1704, and during the rest of the war it rarely rose above 35.) It is easy to understand why the crisis of 1701 should have had a favourable effect on sterling. For the seizure of Antwerp was a threat to the Dutch, and Amsterdam houses may have sought sanctuary for their capital in London. But such considerations can hardly explain the sharp upward turn of the exchanges at the time of the rebellions of 1715 and 1745. In the second of these the value of sterling rose steeply from 34-10, when the Young Pretender landed in Scotland in July, to 37-6 when his forces reached Derby in January, 1746. As the rebels retreated the rate on Amsterdam fell. In April, the month of Culloden, it was down to 36-2, and by July to 35-7. Neither rebellion was a matter of purely English concern. A victory of the Stuarts might have had untoward effects on the Low Countries, and London may well have seemed a safer centre for capital than Amsterdam. But the fact that each of these crises was marked by a fall in the price of British Funds may have played a part in the movement. The Amsterdamers were not insensitive to the lure of a high return on investment: if they bought British stock when its price was low they might hope, in addition, for a capital gain when the crisis had passed. But the chief force operated, almost certainly, on Englishmen. Each political crisis led to a panic in London. Merchants and others had an intense desire for liquidity: they sold commodities, Government stock, and any other asset they could lay their hands on, not so much to foreigners, but to anyone they could find to buy them at home. It was the sale of foreign exchange by Englishmen to other Englishmen, in conditions in which would-be sellers were many and would-be buyers few, that led to a fall in its value, or, in other words, to a rise in the value of sterling. This was so in 1733-4 when fears of war led to an acute demand for cash, and similar sharp upward movements occurred in the crises of 1753, 1763, and 1772. The fact that when, in 1778, the French joined the Americans against Britain, rates of exchange moved sharply in favour of this country, is a final proof that it was reactions in London, and not abroad, that were responsible for the phenomenon. In January and February of that year the London-Amsterdam rate was 34-2; by September it touched 36-6, and in March, 1779, it reached a peak of 37-3. The story was repeated in the last war of the century. In May, 1792, the rate was 36-6. In February, 1793, when revolutionary France declared war on Britain, it stood at 38-7, and by April it was above 40. The highest point of the century was reached in October, 1794, after which, with the occupation of Holland by the French, quotations of rates on Amsterdam ceased.[1]

If the behaviour of rates of exchange in political crises seems at first sight paradoxical, their conduct during the later course of the wars conforms closely to expectations. There was much fluctuation. Hopes of triumph and fears of defeat are faithfully reflected in the course of exchange. In January, 1759, the rate was 34-9; by the summer, with news of success at Minden and Quiberon Bay, it reached 36-6; but two years later, about the time of the Family Compact, it was down to 33-10. Such temporary variations apart, the tendency of continued war was to turn the exchanges against Britain. High government expenditure at home raised prices and made sterling less attractive to foreigners. Obstructions to trade reduced the power of British merchants to earn foreign currency. And the need to pay the forces serving overseas, and to make subsidies or loans to allies, increased the demand for foreign money at a time when the supply was shrinking.

Nor can it be said that British governments always showed wisdom in their handling of the problem of their own remittances overseas. It was the custom to assign the task of making these to a single merchant, or small group of merchants, who undertook to supply foreign currency at fixed rates. The usual procedure, it would appear, was to let it be known in the City that an appointment was to be made, and to leave applicants for the contract to state their terms. During the War of the Spanish Succession, in 1704, Godolphin arranged, in this way, that Sir Henry Furnese should furnish bills on Holland at current rates of exchange, in return for a commission of 11s. per cent. Under Walpole's administration, however, the Treasury relied for finance mainly on members of the 'three monied companies'—the Bank of England, and the East India, and South Sea Companies. In 1740 the contract for the supply of money for the Forces in Jamaica

[1] Details of the crises are given in T. S. Ashton. *Economic Fluctuations in England 1700-1800* (1959).

was given to two leading members of the South Sea Company, Peter Burrell and John Bristow, and in the following year the remittance to the troops in Flanders was put into the hands of John Gore and his associates, Galston and Poyntz. There were loud protests from Sir John Barnard and others in the City at the refusal of the Government to offer the contracts to competitive bidding. The rate at which the funds were remitted imposed a burden on both taxpayers and soldiers; and it was argued, with force, that it would have been more economical if, in each case, the transfer of money had been made in coin or bullion.[1]

The fact that the privileged contractor always set the rate at a low level may have had a depressing effect on rates in the open market. However that may be, the strength of the forces pressing on sterling led to extremely low rates when hostilities were at their height.[2] The boom in trade that followed the end of each war was accompanied by a rise, and the subsequent slump by a fall, in the value of sterling. Throughout the century indeed, whether in war or peace, the exchanges moved favourably as economic activity increased, and adversely when production and trade declined. A sharp change from a high to a relatively low level is generally the harbinger of depression.

IV

British currency was convertible at fixed rates into silver and gold. It exchanged at virtually fixed rates for other currencies similarly convertible. Hence the value of money in England must have been largely determined by the supply of the precious metals, and the demand for these, in the whole group of nations linked in this way. The index numbers constructed by Dr. Gilboy and Mrs. Schumpeter and, for the years after 1779, by Professor Silberling, make it possible to speak with some confidence about changes in the value of money in England.[3] The only other country, however, for which an index

[1] This was urged strongly by Bunce, who declared it to be 'the most frugal method', and added, 'I hope we have now got over that silly old prejudice, against sending gold or silver out of the country.' *Ibid.*, p. 33.

[2] The lowest points reached in each of the wars are as follows: August, 1710, 32; April, 1743, 33-4; November, 1760, 31-8; November, 1777, 32-1; February, 1800, 31.

[3] Elizabeth W. Gilboy, 'The Cost of Living and Real Wages in Eighteenth Century England', *Review of Economic Statistics*, vol. XVIII; Elizabeth B. Schumpeter, 'English Prices and Public Finance, 1660-1822', *idem*, vol. XX, pp. 21-37; N. J. Silberling, 'British Prices and Business Cycles, 1779-1850'. *Idem*, Prelim. vol. V Supplement 2, pp. 219-62.

number of prices is available for the eighteenth century is Sweden. Professor Heckscher made a graph in which Swedish prices were set alongside English prices converted into Swedish money at current rates of exchange.[1] But the figures he used for England were heavily weighted with the prices of Swedish goods brought to this country. Hence it is not of much interest to find that the two series move closely together. The absence of accurate data for other countries makes it impossible to consider English prices in an international setting: all that can be done is to call attention to some influences within the nation that bore on the value of sterling.

Anyone who attempts to trace changes in the value of money must first decide on the commodity or commodities to be used as the standard of measurement. If the value of money is measured in terms of wheat its variation over time will be different from what it would have been if iron or cloth or labour had been selected as the standard. And if several commodities are used conjointly the problem of how much weight to attach to each is important. People are apt to talk about the rise or fall in the value of money without asking themselves the pertinent question, Value to whom? If the object is to determine whether, over a period of time, English exports exchanged for more, or less, imports each series must be constructed from the wholesale prices of things that entered into overseas trade. But an index relevant to the interests of merchants is of little use in deciding whether the cost of living of the agricultural labourer rose or fell. Each question calls for its own index number. Failure to take account of this elementary truth has led to confusion and barren controversy.

One of the chief merits of Dr. Gilboy's study of English prices is that the results are presented in three separate series, the first relating to a fairly wide range of consumers' goods, the second to the same exclusive of cereals, and the third to a number of producers' goods. The chief defect is that the figures are drawn mainly from the records of institutions such as colleges and government departments, and not, like those of Silberling, from prices current. Since these bodies usually entered into more or less long contracts for the supply of the things they needed, the index numbers do not bring out clearly changes in market prices in particular years. They do, however, serve to indicate trends. Years of war excepted, all three series show a gradual downward movement from 1700-1 to

[1] Eli F. Heckscher, *Sveriges Ekonomiska Historia*, Diagram XXXVI.

the middle 'thirties. Thereafter there is little change for about twenty-five years. From 1760, however, the movement is gently upward to about 1783, fairly steady during the next decade, and sharply rising from 1792-3 to the end of the century.

These movements may be explained by changes in the supply of money on the one hand, and changes in the demand for money on the other. In the first thirty or forty years of the century much of the silver that came to England (via Spain) from the mines of Mexico was pouring out again to the East Indies. As has already been said, the shortage of coin was acute. The institution of the public debt had created a mass of new securities, and the chartered companies provided more: to some extent these offered a substitute for cash. They were not, it is true, often used as a medium of circulation. But if they had not existed, landowners, merchants, and other well-to-do people would probably have continued to hold large stocks of money, for reasons of security, as had been the custom of their ancestors. There is evidence that merchants, in particular, held East India bonds, which were transferable and could be turned into cash at short notice.[1] But for this alleviation of the pressure of demand for money (to hold rather than to spend) the downward course of prices might have been steeper.

The relatively stable level of prices between the 'thirties and the 'sixties may have been due to some increase in the supply of coin. Gold was taking the place of silver in the currency, and gold was less in demand for export than silver. These years saw the beginnings of the country banks and the creation of provincial notes and bankers' drafts. But it was not until the Seven Years' War that the supply of money began to outrun the demand. The rapid increase of state expenditure in this war, and still more in the war with the American colonies, added to the mass of securities. And though, as the volume of output rose, the demand for money for use in transactions increased, that for hoarding probably declined. For ten years after 1783 the increased output of paper money by the banks, and of inland bills by traders and manufacturers, was matched by a corresponding increase in the volume of transactions. But, with the outbreak of a new and more costly war, heavy government borrowings from the Bank of England, the rapid creation of country banks, and the expansion of the circulation of inland bills,

[1] A. H. John, 'Insurance Investment and the London Money Market of the Eighteenth Century', *Economica* (new ser.), vol. XX, No. 78, p. 141.

brought an increase in the volume of money so great as to force prices sharply upward. As a result of this and other circumstances mentioned above, there was an external drain of gold, and this led to the suspension of cash payments in 1797. Thereafter, no serious attempt was made to curb the expansion of credit. But the story of further inflation falls outside the limits of this book.

This sketch of the movement of prices is unduly simple. There were other forces operating on both demand and supply. And there were many waves and disturbances of the tidal movements. In each period of harvest failure there was a rise not only in the prices of consumers' goods, but also of producers' goods. In each war, the prices of these rose a little more sharply than those of consumers' goods, for the demand for copper, lead, timber, and other naval stores was strong at such times. Generally, however, the series keep closely in line. Dr. Gilboy's index (as well as the many tables of price-relatives constructed under the direction of Lord Beveridge[1]) gives evidence of the remarkable stability of English prices in the eighteenth century. Agricultural prices, it is true, were volatile, but the amplitude of variation of other prices was small. Dr. Gilboy takes 1700-1 as the base: prices in that year are expressed as 100. For consumers' goods (exclusive of cereals) the figure is 81 in 1735-6, 90 in 1761-2, 113 in 1782-3, 117 in 1792-3, and 168 in 1799-1800. Except for the last interval, the change is relatively small. The eighteenth century saw no such violent decline in the value of money as has occurred in the first half of our own unfortunate century. And this stability of the standard of value is, needless to say, one of the reasons for the stability of the institutions and social relations of the period.

Some historians have argued that rises of general prices have been a powerful factor in transferring resources from wage-earners, land-owners, and professional men to traders and industrialists. Until 1793, however, there can have been little or no involuntary deflection of purchasing power. After that year, rising prices must have led to strict economy by annuitants, officials, and others whose incomes were fixed. In 1795 when an Act was passed to conscript for the navy persons without 'some substance sufficient for their support and maintenance', David Macpherson observed:[2]

[1] *Prices and Wages in England*, vol. I.
[2] D. Macpherson, *op. cit.*, vol. IV, p. 340 n.

'The present ruinous, and rapidly progressive, depreciation of money must soon reduce many gentlemen of small fortune, who *lately* had "some substance sufficient for their support and maintenance" to this description of men qualified for the service of the navy—and their wives and children to the description of paupers, objects of the extorted charity of the parish—if funds can be found to enable the parish to support those who used to support the parish.'

Even in the 'nineties, however, there is no reason to think that the incomes of other classes (landowners, farmers, or even wage-earners) lagged far behind prices. The eighteenth century offers little in support of the thesis[1] that forced savings were a major influence in the rise of large-scale industry.

[1] Earl J. Hamilton, 'Prices as a Factor in Business Growth', *Journal of Economic History*, vol. XII, No. 4, pp. 325-49.

CHAPTER SEVEN

Labour

I

NO generalisations are more unsafe than those relating to social classes. The wide diversity of organisation in English agriculture and manufacture was matched by a similar diversity in the conditions and attitudes of the workers. Many of the writers of treatises and pamphlets tended to ignore these differences: they were obsessed with the problem of the paupers, and, since in an age of economic fluctuation independent workers were liable to fall into poverty, there was a tendency to identify them with that sub-stratum of the population which was rarely, if ever, regularly employed. The insecurity of the standard of life of the wage-earner was attributed not to faulty social or economic arrangements, but to defects of personal character.

According to one group of writers, the workers were by nature indolent, improvident, and self-indulgent. They were addicted to ale and spirits, and their wives drank too much tea. They took little or no part in public worship—at least, in the orderly forms established by law. And if they were indifferent to their fate in the hereafter they were equally insensible to their interests in the world of the living. It was the irrationality of the poor, quite as much as their irreligion, that was distressing. They took no thought for the morrow, and when evils came upon them they cast their burden on others. 'A clergyman, an annuitant, an officer in the army, a tide-waiter or an excise officer, whose income is only seventy pounds a year, is not considered an object of the Poor's Law: they support themselves and bring up their families in credit. Why then are mechanics with equal incomes, less creditable, less independent?'[1] The answer given by some was that the workers were a different order of being. 'I always consider this class of people as in some respect in a state of minority', wrote Charles Townshend. 'They can never from their situation in life, and from the nature of their education, acquire at

[1] T. Jarrold, *A Letter to Samuel Whitbread, Esq., M.P., on the subject of the Poor's Law* (1807), p. 24.

any point in life, such a degree of discretion, as to be capable of judging properly for themselves, or to know without continued guidance, wherein their well-being consists.' They needed discipline. 'To hold the lower orders to industry and guard the morals of the poor', another writer declared, 'is the truest patriotism.'[1]

Some observers held that the trouble arose from the payment of unduly high wages. In 1739 William Temple asserted that the only way to make the workers temperate and industrious was 'to lay them under the necessity of labouring all the time they can spare from rest and sleep, in order to procure the common necessaries of life'. And Arthur Young declared that 'everyone but an idiot knows that the lower classes must be kept poor or they will never be industrious'. Enclosure was desirable if only because rights of common led to irregularity of work.[2] The subsidy on the export of grain was a good thing because it raised the price of bread and so forced the poor to work harder and better. 'The best goods are made in the worst times. When employment is scarce, every manufacturer strives after perfection in his particular branch, for fear of being dismissed. In dearness of provisions it is just the same. If wheat sells for 10 or 12s. a bushel, the manufacturers are obliged to work more and debauch less.'[3] High taxation of gin and ale appeared to be the obvious remedy for drunkenness. The suppression of fairs and wakes, and of the interludes and other diversions that went with them, might prevent extravagance and abstention from labour. And there was much to be said for the payment of wages in goods (of the right kind) instead of in money.[4]

Temple was a clothier, embittered by a conflict with the turbulent weavers of Gloucestershire. Neither he nor the other publicists who wrote in the same strain took account of the setting of domestic production. Employment in manufacture was seasonal. In the rural areas the workers were called on to lend a hand in making hay, reaping corn, picking hops and fruit, and transporting and storing the crops. From July to October the weavers were away from their looms. In Cornwall men divided their time between mining and fishing for pilchards;[5] in Yorkshire the fulling mills often worked

[1] *Reflections on Various Subjects*, pp. 62-3.
[2] See E. Gilboy, *op. cit.*, p. 144.
[3] William Temple, *The Case between the Clothiers and the Weavers*, reproduced by J. Smith, *Memoirs of Wool*, vol. I, p. 30.
[4] For citations of such views see E. S. Furniss, *The Position of the Labourer in a System of Nationalism*, p. 126.
[5] J. Rowe, *op. cit.*, p. 28.

Labour

only between October and Whitsuntide, when water was plentiful;[1] and occasionally industrial as well as agricultural labourers had to down tools to work on the roads. Their own specialised employments, moreover, were often subject to interruption by a delay in the receipt of material, or a temporary fall in the demand for the product; and a storm or a pestilence sometimes put an end to traffic on the roads, and so led to involuntary idleness.

Apart from such influences, however, the normal condition of most domestic producers was one of under-employment. Each master manufacturer liked to have at his disposal a number of workers in excess of his need in ordinary times, so that in periods of brisk trade he would not be hampered by shortage of labour. The possibility of working on their own scraps of land, of obtaining jobs on the farms, and (in the case of women spinners) of falling back on the earnings of other members of the household, led the workers to acquiesce in the arrangement. Hence there existed at many points of the economy a pool of labour similar to that at the docks in our own day. More men and women were attached to each industry than could normally find full-time employment in it: the surplus of labour to which many writers called attention at various times was made up less of men permanently out of work than of those whose hold on employment was precarious. Nor was this state of things peculiar to domestic industry. It existed in the towns and especially in London. Shipping and river-side occupations were notorious for irregularity. The Capital attracted labour from all parts of the country, and it was not possible quickly to adjust demand to supply. At the end of the century it was said that there were seldom fewer then ten thousand domestic servants out of a place in London.[2]

It would be wrong to imagine that what the ordinary man sought above all was continuity of work. Casual methods of hiring naturally engendered casual habits. It is true that the workers resented the imposition of unemployment. 'Such is the manner of life (from hand to mouth) and the peculiar improvidence of labouring manufacturers and mechanics, beyond those in plain, simple husbandry that as often as there happens an epidemical sickness, or a scarcity and dearth of provisions, or a rigorous season, to put a stop to their work and wages, so often, be the general state of trade and manufacture never so good, will there be great occasion of complaint

[1] W. Crump and G. Ghorbal, *op. cit.*, p. 16.
[2] By Patrick Colquhoun cited by M. D. George, *op. cit.*, p. 112.

among that class of people; and numbers of them will become objects of relief.'[1] But leisure, at times of their own choice, stood high on their scale of preferences. If it is true that they observed fewer holy days than their fellows of Papist France, it would be idle to deny that a large part of their energies went into channels largely outside the economic system. In most parts of the country there were annual fairs or wakes during which little or no work was done. The journeymen of London downed tools on each of the eight hanging days at Tyburn, to say nothing of the frequent occasions when they joined in a riot or demonstration.[2] A contest in the ring or a race-meeting often led to a cessation of labour. In 1754 the colliers of Oldham and the weavers of Manchester competed in a cock-fight that lasted three days.[3] Coal miners left the pit to take part in a Parliamentary election, to celebrate a national victory, or, as in 1779, to seek consolation for a national defeat.[4]

Such occasional abstentions from labour were, however, of small account. Far more serious was the almost universal practice of working a short week. 'It is not those who are absolutely idle that injure the public', remarked a writer in 1752, 'so much as they who work but half their time.'[5] A clergyman asserted that few of the working people of Manchester were regularly employed above two-thirds of the week; and Eden declared that the labourers of the same town rarely worked on Mondays and that many of them kept holiday for two days more.[6] No doubt upper-class observers were apt to exaggerate, but there can be little doubt about the tendency to lengthen the Sabbath. It was a working cutler, writing for the delectation of his fellows, who described

> How upon a good Saint Monday
> Sitting by the smithy fire,
> Telling what's been done o' t' Sunday

he was interrupted by the sudden appearance of his wife, who called him a great, fat, idle devil, threw his liquor in his face, and concluded with a threat:

> 'od burn thee, Jack, forsake thy barrel
> or never more thou'st lie wi' me.

---

[1] John Smith, *op. cit.*, vol. I, p. 390.
[2] M. D. George, *op. cit.*, p. 208.
[3] A. P. Wadsworth and J. de L. Mann, *op. cit.*, p. 391.
[4] T. S. Ashton and J. Sykes, *op. cit.*, p. 167.
[5] *Reflections on Various Subjects*, p. 70.
[6] A. P. Wadsworth and J. de L. Mann, *op. cit.*, p. 388.

But the rhyme of *The Jovial Cutler*[1] brings out one aspect not mentioned by the censorious publicists. The wife's tongue is described as moving more quickly than the cutler's boring stick 'at a Friday's pace'. If there was idleness at the beginning of the week there was often extreme pressure towards the end of it. Possibly if the workers had been willing to behave on the other five days as they did on Fridays both the national income and the standard of comfort of the labouring classes would have been higher. But Adam Smith thought it unlikely:

'Excessive application during four days of the week is frequently the real cause of the idleness of the other three, so much and so loudly complained of. Great labour, either of mind or body, continued for several days together, is in most men naturally followed by a great desire for relaxation. . . . It is the call of nature, which requires to be relieved by some indulgence, sometimes of ease only, but sometimes, too, of dissipation and diversion.'[2]

He was not alone in this opinion. Some years before he wrote, the shrewd Malachy Postlethwayt had expressed the view that the high quality of English manufacture was to be attributed to the frequent 'relaxation of the people in their own way'.[3]

It might be thought that so long as industry rested on household production the distribution of time between work and leisure was no concern of the employer: all that mattered to him was that the finished goods should be delivered at the end of the week or fortnight prescribed by custom. But over long periods, the worker's decision as to leisure seemed to some writers to exhibit a perversity that did harm to both the employer and the public. 'It is observed by clothiers and others, who employ great numbers of people', wrote Petty, 'that when corn is extremely plentiful the labour of the poor is proportionately dear: and scarce to be had at all (so licentious are they who labour only to eat or rather to drink).'[4] As has been pointed out, a low price of corn was conducive to a high demand for manufactured goods. It was precisely when the master offered

---

[1] It is believed that the author was a man known as Old Bone-heft. The verses are printed in *The Songs of Joseph Mather*, pp. 189-90.
[2] *The Wealth of Nations*, vol. I, p. 73.
[3] Cited by E. S. Furniss, *op. cit.*, p. 126.
[4] Sir William Petty, *Political Arithmetic* (Hull's ed.), vol. I, p. 274. Cited by T. E. Gregory, 'The Economics of Employment in England, 1660-1713', *Economica*, vol. I, no. 1, p. 274.

more employment that the worker (content, it was said, with a fixed standard of consumption) chose to enjoy more leisure. When dearth came and the demand for labour was reduced, men clamoured for employment and were ready to work the whole of the week: a manufacturer told David Hume that, during the famine of 1740, his employees not only made shift to live, but actually repaid debts contracted in earlier years. But, here again, Adam Smith dissented.[1]

Social history has sometimes been written in terms of debtors and creditors, on the assumption that the two belonged to different social classes. The creditor is supposed to be rich, the debtor poor. English society was shot through with credit, but the lines ran in various directions. A landlord might provide a tenant with much of the fixed capital of the farm and sometimes also with part of the working capital. He might, like the proprietor of Thornborough in the North Riding, make advances to a small farmer,[2] or, like Lord Dartmouth, lay out money on fulling mills and lease these to tenants on his estate.[3] But if he were eager to develop his properties he might borrow, as the Duke of Bridgewater did, from his tenants. A merchant might finance a manufacturer, but, equally, a manufacturer might supply goods to a merchant on credit. Most entrepreneurs were, at one and the same time, borrowers and lenders, debtors and creditors. Credit penetrated to relationships from which it is largely excluded today, and often it was humble people who were forced to provide it. Farm servants normally received board and lodgings but had to wait until the end of the year for a payment in cash. Men serving in the Army were supposed to get their pay (at the rate of 8d. a day) every two months.[4] But paymasters were often behindhand. Seamen in the Navy were usually given an advance of two months' pay when they first went to sea; but it was believed they would be less prone to desert the ship if there was money owing to them, and hence nothing more was given till the crew

---

[1] 'That a little more plenty than ordinary may render some workmen idle, cannot well be doubted; but that it should have this effect on the greater part, or that men in general should work better when they are ill fed than when they are well fed, when they are disheartened than when they are in good spirits, when they are frequently sick than when they are generally in good health, seems not very probable. Years of dearth, it is to be observed, are generally among the common people years of sickness and mortality, which cannot fail to diminish the produce of their industry.' *The Wealth of Nations*, vol. I, p. 74.

[2] E. Gilboy, *op. cit.*, p. 152.

[3] W. Crump and G. Ghorbal, *op. cit.*, p. 72.

[4] *The Necessity of Establishing a Regular Method for the Punctual Payment of Seamen in the Royal Navy* (1758), p. 7.

was paid off. It is said to have been a great boon to the seamen when, during the Seven Years' War, some regular provision was made for their wives and families, and the balance due to them was handed over every six months.[1]

In manufacture somewhat similar conditions prevailed. Work was given out and returned at irregular intervals. The employer was accustomed to make up his accounts with his workers, not weekly, but at the end of a period that might run to two or three months: when the worker was able and content to wait from one settlement day to the next he was, in effect, giving credit to his employer. Few, however, could afford to do this. The master was called on to make interim payments for subsistence—whether in money or goods—and when the reckoning was made at the end of the period, far more often than not the worker found himself in debt. It is impossible to appreciate the conditions of the workers in the eighteenth century without taking account of this fact. In a petition to Parliament in 1752, it was asserted (no doubt with some exaggeration) that there were over 20,000 'working manufacturers' in Birmingham who were continually contracting small debts which their creditors found difficulty in recovering.[2] Wherever small-scale, domestic production existed, the worker was bound to his employer, not by contract but by the fact that he could not transfer to another until he had paid what he owed. His concern was with the immediate situation: his food for the next few days, the money needed to tide over a domestic crisis, the means of satisfying the claims of a shopkeeper whose patience was exhausted. He was more apt to seek a loan than an increase of wages. His problems were, or seemed to be, personal rather than collective. So long as the wage contract was mixed up with a loan contract, each was, almost of necessity, arrived at by individual bargaining.

One reason for the long pay and the indebtedness associated with it was the shortage of coin, especially that of small denomination. Employers of labour, and even banks, had to pay suit to retailers, turnpike-keepers, and others whose businesses enabled them to accumulate stocks of small change. Some industrialists sent their workers to the ale-house to draw their wages, and others who paid directly in coin did so only at infrequent intervals and in coins of large

---

[1] Lieut. Robert Tomlinson, 'Essay on Manning the Royal Navy', *Pub. Navy Records Society LXXIV*, p. 121n.
[2] *J. H. C.*, vol. XXVI, p. 414. Cited by H. Hamilton, *The English Copper and Brass Industries to 1800*, p. 273.

denominations. The worker who was paid in this way often visited the inn to turn the money into small change—or at least, no doubt, this was how he explained things to his wife. Yet other employers, like Samuel Oldknow, made payment in notes drawn on shopkeepers, or, like the Strutts, the Peels and the Gregs, had truck shops of their own where part, at least, of the wage was handed over in the form of goods. Whatever the makeshift, it bore heavily on poor men. They had to pay dearly for the things they acquired. Choice was restricted. They were unable to buy goods in small quantities. And thrift (which is proverbially a matter of caring for pennies) was made difficult.

In some industries there was a tradition that the workers should have a share of the product of their labour. Both the coal-hewers of the north and the coal-meters of the Thames received by custom an allowance of fuel; and ironworks and other establishments that used coal often supplied it on special terms to their workers. The mates of the West Indiamen had a right to the sweepings of sugar and coffee from the hold of the ship; the gangsmen and coopers established a claim to the drainings of molasses and spilt sugar on the floor of the warehouse; and the labourers in the corn ships believed themselves to be similarly entitled to the grain that had been removed as samples. At the Royal Yards, the shipwrights were allowed to take for firewood the chips that fell from the axe, and their womenfolk were permitted to do the gleaning. In each case the workers saw to it that the crumbs from the master's table were ample. Casks were handled not too gently; sacks were liable to burst open; shipwrights took care that their wives did not go short of firewood. The line of demarcation between the extension of established rights and barefaced robbery is difficult to draw. The London coal-meters accepted 'winking money' from the merchants in return for giving over-measure,[1] and the banksmen and wagoners at the Whitehaven collieries were similarly accused of accepting bribes from shipmasters.[2] If the dock and riverside crafts lent themselves peculiarly to pilferage—the bum-boat men were notorious for their traffic in stolen goods—the practice was also widespread in manufacture. Yarn returned to the warehouse might be of light weight: the deficiency was hidden by moistening with water, oil, or butter.

[1] T. S. Ashton and J. Sykes, *op. cit.*, p. 209.
[2] O. Wood, 'A Cumberland Colliery during the Napoleonic War', *Economica*, vol. XXI, no. 81.

Wool pickers, spinners, and weavers had ample opportunities of purloining material and of disposing of it by sale. In the hosiery industry there were, it was said, no fewer than 120 'Turkey merchants' who traded in stolen yarn.[1] In the northern coalfield the hewers and barrowmen were accused of putting large coals in the corf (or basket) in such a way as to leave the bottom empty, covering the large pieces with small coals at the top.[2] Industrial concerns had to take special precautions to avoid losses of fuel from the wagons and from the stocks at the works. The Coalbrookdale Company set aside a supply of coal for sale to their employees. But on March 26th, 1796, the manager, Richard Dearman, issued a statement that 'many People have of late feloniously taken Coals from off the Hearths and out of the waggons and Crates, also from the Incline-plane, and that some Workmen carry away Coals covered with Cinders or small Brays, others cut Ends off Boards and other useful Stuff into Chips and carry the same Home, and some Waggoners sell or give Coals from off the Waggons, on the Road betwixt the Pits and the Coal-Hearth.'[3]

That there was a close connection between the 'long pay' and embezzlement is evident. If the employer were dilatory in paying for work or niggardly in making advances, the worker fell in debt to the butcher, the baker, and the innkeeper. The poor, like their betters, had the habit of buying on credit: retailers had to wait long periods for their money, and the incidence of bankruptcy among shopkeepers was high. 'They rarely receive money for what they sell', wrote Charles Townshend, 'and being as backward in their payments for what they owe, find a quick transition from their shops to a prison.' It was not always easy for the wage-earner to get extended credit unless pledge of payment were given. Goods purloined might serve this purpose. 'Last week the Magistrates of Kidderminster convicted a Publican's Wife within their jurisdiction, for having in her Custody, a certain Quantity of Worsted Yarn, which she had received by way of Pledge, for two shillings borrowed of her by a Journeyman Weaver's Wife.'[4] Notices like this in a local

---

[1] W. Felkin, *A History of the Machine Wrought Hosiery*. Cited by E. Lipson, *An Economic History of England*, vol. II, p. 61.

[2] T. S. Ashton and J. Sykes, *op. cit.*, p. 19.

[3] It was announced that any employee found guilty should suffer a fine of 5s., and that outsiders charged with such offences should be prosecuted. *MS. Records of the Coalbrookdale Co.*

[4] *Aris's Birmingham Gazette*, June 29th, 1778,

newspaper throw a glimmer of light on the finance of an underworld that has left few other records. If the shopkeeper or the innkeeper were over-scrupulous the worker who lived in a town might have resort to the pawnshop, an institution that played a large, and not always dishonourable, part in the lives of the poor in this century. Pawnbrokers were accused of charging excessive rates for the 'low credit' they provided. But the chief count against them was that they encouraged petty theft and, in particular, the embezzlement of the materials of industry. In defence, one of them pointed out that in the case of customers belonging to occupations such as those of the tailor, watchmaker, mantua maker, and laundress, it was impossible to distinguish master or mistress from servant. But he also pointed out that 'it would . . . be a great means of preventing this mischief if master-employers and others would always pay poor manufacturers . . . as soon as their work is done.'[1]

II

The process by which gradually (and sometimes, no doubt, painfully) the working classes adjusted their ways of life to the needs of expanding manufacture is not easy to trace. Pressures of various kinds played a part. A survey of the measures passed to suppress embezzlement and delay in returning materials shows a progressive increase of penalties. Under an Act of 1703 (relating to the woollen, linen, fustian, cotton, and iron industries) a worker found guilty of purloining goods entrusted to him had to forfeit twice their value. Under another of 1740 the cost of the prosecution was added to the penalty. Nine years later what had previously been treated as a breach of contract was made a criminal offence, punishable by fourteen days' imprisonment. In 1777 the term was increased to three months, and, for a second offence, to six months. The Act of 1749 required that work put out should be returned within twenty-one days: in 1777 the period was shortened to eight days.

Administration, however, was weak. According to Macpherson, manufacturers were 'remiss or unwilling to expose themselves individually to the revenge of the delinquents in punishing breaches of

---

[1] *The Business of Pawn-Broking stated and defended.* By a Pawn-broker (1744), p. 55.

Labour 211

the law against frauds'.[1] Hence, in the woollen and worsted industry, corporate action was taken. In a large number of counties the employers set up committees (with fifteen or so members) to administer a fund subscribed for the prosecution of workers charged with such offences as embezzlement, defective spinning, wetting or oiling cloth, and delay in returning materials. The inspectors appointed by these bodies were licensed by the Justices of the Peace, and so formed what has been termed an industrial police.[2] In many towns employers in different trades established societies 'for the prosecution of felons'. Actions relating to indebtedness could be brought in the Court Leet, but in some industrial areas special courts for the recovery of small debts were set up: it became increasingly difficult for a man who owed money to one employer to find work with another.

If the conditions of domestic production called for industrial discipline those of the workshop, factory, or the mine did so in greater measure. When the worker was a member of a team and made use of appliances belonging to an employer, his absence might reduce the output of his fellows and mean that part of the plant stood idle. 'I have not half my people come to work today', wrote Lewis Paul from his London establishment in 1742, 'and I have no great fascination in the prospect I have to put myself in the power of such people.' One remedy, widely adopted, especially in the case of skilled workers, was to impose a contract for a period of service, in which the hours and conditions of labour were set forth. Any breach of the contract might result in an action at law. In the northern coalmines the period of the bond was a year, or eleven and a half months, and by the end of the century an elaborate code of rules was incorporated in the document. In metal-mining, iron smelting, glassmaking, and the production of paper and chemicals, the skilled workers were all required to make contracts for periods of service ranging from three months to seven years or more. Whenever an employer in a new trade had spent time and effort in training a man for a particular purpose he took care to ensure against losing his services to a competitor: all the engine-erectors employed by Boulton and Watt were bound to the concern for long periods.

A contract of service did not give a guarantee of complete regularity of attendance. Whatever the document might say, coal-miners

---

[1] Macpherson, *op. cit.*, vol. IV, pp. 44, 72.
[2] H. Heaton, *The Yorkshire Woollen and Worsted Industries*, p. 423.

in all parts of the country continued to absent themselves from the pits when they felt inclined, and the letters of employers in other industries give evidence of much voluntary unemployment. But at least the worst offenders could be so dealt with as to deter the rest from going too far. In addition to contracts with individual workers, some of the larger employers issued regulations as to hours of labour and gave notice of penalties for breaches of discipline.

In the early years of the century Ambrose Crowley required his men to labour for thirteen and a half hours on each of the six working days. Towards the end of it, the 'Rules for the Preservation of good Order in the Works of Wm. Reynolds & Co.' prescribed a twelve-hour day from 6 a.m. to 6 p.m. (including an hour and a half for meals). Any workman who failed to keep to the times laid down was subject to a fine of a shilling; if he earned twelve shillings or more a week the penalty was increased to half-a-crown. Those who lingered unduly over their meals had to forfeit a quarter of a day's wage. And any man who left his employment without having given a month's notice at the previous reckoning was to forfeit his wages up to, but not exceeding, 10s. 6d. If production was still frequently interrupted by the passion of the worker for revelry the master had power to retaliate. 'We are laying by for Xmas at our works', wrote Josiah Wedgwood to his partner, Bentley, on December 31st, 1772. 'The men murmur at the thoughts of play these hard times, but they can keep wake after wake in the summer when it is their own good will and pleasure and they must now make a few holidays for our convenience.'[1] Pressure of this kind, no doubt, had some effect. Little by little the workers learnt to conform to conditions laid down in their contracts of employment and became inured to regularity of work. The process was long and can hardly have been pleasant. But there is no need to speak of it indignantly as the 'conditioning of labour' or to attribute it simply to capitalist exploitation. Today 'discipline' has become a word of ill-repute. Yet it is obvious that if the workers of the eighteenth century had refused to conform to some code of conduct when at work, there could have been no factory system, and no such rise of output, and hence of the standard of life, as was, in fact, attained in the nineteenth century.

It was not, however, simply the fear of penalties that effected the

---

[1] *Letters of Josiah Wedgwood to Bentley*, vol. 2, p. 14. I am indebted for this reference to Dr. Reginald Moss.

change of attitude to work. Not all economists, nor all employers, shared the views of Temple and Young. At least one publicist protested against a code of morality that imposed a more austere standard on the poor than on the rich.[1] And Adam Smith insisted that high wages were to be regarded not merely as a burden but also as an incentive to output:

'The liberal reward of labour', he wrote, 'as it encourages the propagation, so it increases the industry of the common people. The wages of labour are the encouragement of industry, which, like every other human quality, improves in proportion to the encouragement it receives. A plentiful subsistence increases the bodily strength of the labourer, and the comfortable hope of bettering his condition, and of ending his days in ease and plenty, animates him to exert that strength to the utmost. Where wages are high, accordingly, we shall always find the workmen more active, diligent, and expeditious than where they are low: in England, for example, than in Scotland; in the neighbourhood of great towns than in remote country places.'

As early as 1755 Bishop Berkeley had pointed to another approach to the problem when he asked 'whether the creation of wants be not the likeliest way to produce industry in a people?' There was a limit to the quantity of locally produced bread, meal, or bacon that a family would wish to consume—and an increased supply of ale afforded no guarantee of increased output. If the workers were to be induced to set less store by leisure it was necessary, in the first place, that they should be paid in freely exchangeable currency. Recounting the part played by the first Abraham Darby in creating a new community at Coalbrookdale, his daughter-in-law wrote: 'This place and its environs was very barren, little money stirring amongst the inhabitants. So that I have heard they were obliged to exchange their produce one to another instead of money, until he came and got the works to bear, and made money circulate among the different parties who were employed by him'. Something has already been said about the shortage of money in the country districts, and especially in those, like the region Abiah Darby spoke of, remote from large towns. It was plain to industrialists that if they were to succeed they must put purchasing power at the disposal of their workers:

---

[1] 'Tippling in an ale-house may be punished, but not drinking in a tavern; bawdy-houses may be searched but not bagnio's and so in every other instance the laws themselves vindicate our tyranny over the poor.' *Reflections*, p. 72.

coin of the realm if possible, but, failing that, some home-made form of currency.

But money alone would not solve the problem. There had to be also a wider variety of goods than could be produced locally. Richard Arkwright has been praised—perhaps unduly—for his gifts as an inventor. His real claim to fame rests on his skill in co-ordinating production and his powers as an organiser of labour. It is unlikely that he ever read the works of Berkeley, but he reached the same conclusion as to the necessity of arousing and satisfying new wants. In 1778 he had adopted the practice of giving 'distinguishing dresses' to the most deserving of his workers, 'which excites great emulation', and arranged balls at the Greyhound Inn at Cromford, where no doubt these were displayed.[1] In a manuscript diary kept by the Hon. John Byng, who paid two visits to Cromford in 1789 and 1790, the following entry describes other steps taken later by Arkwright:[2]

June 18, 1790. . . . 'At two o'clock I was at the black dog at Cromford; around which there is much levelling of Ground and increase of Buildings for their new Market (for this place is now so populous as not to do without) which has already been once held and will be again tomorrow. . . .

'The Landlord has under his Care a grand Assortment of Prizes from Sir R. Arkwright, to be given at the years end and to such Bakers, Butchers, etc., as shall have best furnished the Market. How this will be peacably settled I cannot tell!! They consist of Beds, Presses, Clocks, Chairs, etc., and bespeak Sir R'd's Prudence and Cunning, for without ready Provisions his Colony could not prosper. So the Clocks will go very well.'

The parallel growth of production, the supply of money, and opportunities for consumption, may be illustrated by the story of the Cheshire village of Wilmslow, twelve miles or so south of Manchester. In his unpublished history of the parish, written in 1785, Samuel Finney describes how, forty years earlier, the chief occupation, other than agriculture, had been the making of mohair and silk buttons. Two manufacturers from Macclesfield came weekly to

[1] *Derby Mercury*, September 25th, 1778. I am indebted to Dr. R. S. Fitton for this reference.
[2] See article on the Arkwright Bi-centenary. 'The "Cotton Lord" at Home. A Factory Village.' By A. P. W., *Manchester Guardian*, December 23rd, 1933. The diary of John Byng, afterwards Lord Torrington, is in the Municipal Central Library, Manchester.

put out material and take in the finished product. They employed mainly women and children and, between them, paid about £25 a week to the villagers, the more skilled of whom might receive about 3s. 6d. for a week's work. Trade was confined to a few petty retailers who supplied 'treacle, brown sugar, salt, tobacco, coarse linens and woollens, and other small necessaries'. Finney could think of only two shoemakers, but there were about a dozen clog-makers who made use of leather from old shoes brought in from neighbouring towns. The diet of the poor consisted of 'barley bread, potatoes, buttermilk, whey and sower porridge made with water, crabb juice and a little oatmeal mixed and boiled and sweetened with a little treacle'.

When a change in fashion in favour of metal buttons led to a decline of the staple industry, the place of the Macclesfield employers was taken by Yorkshire clothiers, who put out the spinning of wool on Jersey wheels.[1] Finney declared that the advent of Yorkshire enterprise was 'one of the most favourable events that had ever happened to the inhabitants'. Within a few years, 'there were few houses, the farmer's not excepted, which had not wheels a-going in them', and about £50 a week was being paid out in wages. Children of six years of age 'could almost earn their living'; a child of eight might make 3d. or 4d. a day, and a diligent woman 4s. a week. Later, in the 'seventies, a new stream of enterprise, springing from Stockport, completely altered the life of the parish. Cotton took the place of wool, and the jenny drove out the simpler Jersey wheel. Before long, Wilmslow men themselves were setting their fellows to spin yarn for the making of coarse calicoes. Attracted by the possibility of earnings that ranged from 7s. to 10s. a week, 'women forsook their wheels, the men their ploughs, spades and flails to learn this nice Art'; and not more than about twenty Jersey spinners remained in the village. Most of the work was still done in the home. But a local capitalist, Bower, had recently built near the church in the centre of the village, a small factory with a water-wheel, where the preliminary processes of carding and slubbing were concentrated. (Difficulties arising from an insufficient supply of money were overcome, it is said, by the device of paying wages at intervals of a few

---

[1] The Jersey wheel was sometimes called the 'big-wheel'. It was turned by hand in one direction to give a twist to the roving, then in the other to wind the thread on to the bobbin. It was a less efficient device than the Saxony wheel which was worked by a treadle, twisted and wound continuously, and could spin two threads at the same time. W. B. Crump and G. Ghorbal, *op. cit.*, pp. 35-40.

hours, so that there was time to recover from the shopkeepers the money spent by a first group of wage-earners, before the next group appeared to draw their earnings.) Finally, in the 'eighties, a capitalist, Samuel Greg, built, a mile and a half away at Styal, a large mill, worked by water-power, for carding, slubbing, and spinning yarn for warps to be used in the weaving of fine muslins.[1] According to Finney, the wages paid here were 10s.-12s. for men, 5s. for women, and 1s. to 3s. for children. In a little over forty years Wilmslow had turned from buttons to woollen yarn, from this to coarse cotton weft, and from this, in turn, to fine cotton warp. For some, employment in the factory had taken the place of work in the home. Earnings had risen substantially and the flow of money had quickened.

The process had involved an expansion of retailing. By 1785 the number of shopkeepers had 'increased amazingly, some of whom dealt in a great variety of articles . . . tea, coffee, loafe sugar, spices, printed cottons, calicoes, lawns, fine linens, silks, velvets, silk waistcoat pieces, silk cloaks, hats, bonnets, shawls, laced caps and a variety of other things'. Subsidiary employments had come into being. The number of joiners, carpenters, brickmakers, and bricklayers had greatly increased. There were now a dozen shoemakers and 'perhaps not above two clog-makers'. Diet also had been transformed: 'though they continue constant to their potatoes, they have utterly forsaken all the rest and use fresh meat or bacon, the fatter the better, wheat bread, and that generally of the finest, well buttered, and tea, forsooth, generally thrice a day.' The picture is the more striking in that it is painted by one who was no friend to technological change: it stands in sharp contrast to that presented by Temple and others of a body of labour, supine, unambitious, and insusceptible to the offer of improved standards of life.

That the development of industry was associated with the growth of foresight and thrift is attested by the rise, towards the end of the century, of institutions such as friendly societies, trade clubs, savings banks, and building societies. That it stimulated, and aided the achievement of, personal ambition is shown by the number of men who began life as wage-earners and ended it as employers. The emergence of an independent, self-respecting class of wage-earners was the result of the interaction of many forces: the ending of the Gin Age, the spread of nonconformity with its stress on personal

[1] For further details see Francis Collier. *The Family Economy of the Working Classes in the Cotton Industry, 1784-1833*. Chapter V (1965).

Labour

responsibility (and notably by the preaching of Wesley), the growth of education through charity schools and other institutions, the spontaneous rise of workers' organisations, and so on. Employers played a part, both by the enforcement of industrial discipline and by benevolent action. But in the main, it would appear, the social changes of the century derived from economic processes: on the side of production from an increased division of labour, an extended application of capital, and a more closely knit organisation; on the side of consumption from a widening of the range of commodities for sale, and a quickening of the desire to enjoy a higher standard of comfort. That expenditure was not always such as the well-to-do approved is a minor matter. It is open to question whether, as time went on, the workers became more or less sober. A nineteenth-century observer, looking back, was of the opinion that there had been little improvement here. He remarked that the workers now lost less time in drinking by days together, but there was more drinking at night and at the end of the week. In this respect, as in others, habits had conformed to the requirements of industry.

III

The wage-earners have never been a homogeneous group. In the eighteenth century some, like the masons, cabinetmakers, and smiths, owned their workshops and tools. Others, like the frame-work knitters, paid a rent for their appliances. Others again, too numerous to specify, worked with instruments owned and controlled by their employers. Some directed, and paid out of their earnings, the apprentices, journeymen, or women who assisted them. The pitman had his marrow, the smith his striker or hammerman. Sawyers worked in pairs (the top-sawyer at ground level, the under-sawyer in the dust of the pit) and received a joint daily wage. Sub-contracting and collective payment were widespread. Craftsmen and mechanics were paid by the hour, the day, or the week, but the earnings of textile and metal workers, along with many others, varied with their output. 'Almost all master manufacturers now find it to their interest to pay their workpeople by the piece or the great', wrote a plagiarist of Dean Tucker in 1757.[1] Some wage-earners had a daily task to perform for a fixed number of hours; others worked for periods of their own choosing. Some were paid in cash; others received part of their wages in goods. In agriculture and coal-mining it was common for

[1] 'Sir John Nickoll', op. cit., p. 19.

the employer to provide his worker with a cottage, free of charge
or at a low rent. There were perquisites of varied kinds. In 1758
a farm servant at Thornborough was allowed half a guinea for a hat
and washing, and a guinea because of 'having no occasion for to
buy him a frock'.[1] The keelmen of the Tyne, like the labourers in
the field at harvest time, received a customary allowance of bread
and beer when at work.[2] The pitman had a monthly supply of coal,
either free of charge or at a low price. In the Midland coalfields
employers sometimes provided their colliers with beef, and it was
the practice to compensate for work of a disagreeable or dangerous
nature, by an allowance of beer or ale. On the other hand, wages
were subject to deductions not only for defective work, but also
for things supplied by the employer: picks and explosives for the
copper-miners, shovels for the London coal-heavers, candles for
weavers and knitters, and so on. At the beginning of the century
deductions were made at Ambrose Crowley's works to pay for the
services of the surgeon and chaplain; and in the seventeen-nineties
at the Horsehay ironworks (to take one of several similar instances)
the workers had to contribute 6d. a head each month to the doctor.

In some industries (of which agriculture was the chief) justices
of the peace occasionally still performed their statutory duty of regulating rates of wages and hours of labour. But over most of the
country no assessments were made. When in 1756 the weavers of
Gloucestershire complained of low wages and petitioned for relief,
the only remedy that occurred to legislators was to prohibit truck
and lay down rules for the assessment of wages by the justices.
But by this time influential opinion was strongly opposed to state
intervention in industrial relations, and in the following year Parliament authorised free bargaining.[3] Nevertheless, whenever the peace
of the metropolis itself was threatened by disputes about wages
it tended to fall back on the traditional system of regulation. Under
an Act of 1721 a maximum had been set to the wages of the journeymen tailors, and the justices continued to enforce this.[4] In 1770,
perhaps because of the disorders connected with John Wilkes, an
Act was passed to fix the wages of the unruly coal-heavers.[5] And,

---

[1] E. W. Gilboy, *Wages in Eighteenth Century England*, p. 157.
[2] T. S. Ashton and J. Sykes, *op. cit.*, p. 196.
[3] E. Lipson, *The Economic History of England*, vol. III, pp. 266-7.
[4] *Ibid.*, p. 405.
[5] T. S. Ashton and J. Sykes, *op. cit.*, pp. 205-6. The measure was allowed to lapse after three years.

three years later, when troubles in the silk industry were acute, Parliament imposed on the Lord Mayor and aldermen of London and the justices of Middlesex the obligation of assessing the wages of the weavers, under what came to be known as the Spitalfields Act. Generally, however, the determination of wages was a matter for bargaining between masters and men. As will be seen, custom played a large part, and the tradition of apprenticeship had an influence on both supply and demand. The market for labour was far from perfect. In agricultural and mining areas the workers attended the annual fairs to offer themselves for hire; and Spitalfields market served as an employment exchange for the poor women silk weavers of London.[1] Employers sometimes gave notice of vacancies in local newspapers, and in cases of urgency made use of the town crier. The office of the overseer of the poor, the inn, and the house of call of the tramping artisan, all served as centres of information. Most often, however, it was by word of mouth from one man to another that news of opportunities of employment was spread. On the side of the masters, at least, the area of competition was relatively narrow: it is hardly possible to speak of a regional, let alone a national, market for labour.

Most of what can usefully be said about wages was said long ago by Adam Smith.[2] One of his objects was to show that there were no grounds for the belief, held by many contemporaries, that they were determined by what was necessary to maintain a labourer and his family or by 'the lowest rate which is consistent with common humanity'. There were seasonal variations in wages. Because of the greater hardships and hazards involved, seamen usually received higher pay in the winter than in the summer. In many occupations, however, the reverse was the case. 'But', Smith said, 'on account of the extraordinary expense of fuel, the maintenance of a family is most expensive in winter. Wages, therefore, being highest when this expense is lowest, it seems evident that they are not regulated by what is necessary for this expense; but by the quantity and supposed value of the work.'[3] He went on to point out that wages did not fluctuate with the price of provisions. 'These vary everywhere from year to year, frequently from month to month. But in many

---

[1] According to Defoe the so-called Mop fairs or Statute fairs, which the workers attended, tools in hand to indicate their callings, were dying out as early as the 'twenties. *Tour*, vol. II, p. 31.
[2] *The Wealth of Nations*, Bk. I, ch. VIII.
[3] *Op. cit.*, p. 65.

places the money price of labour remains uniformly the same sometimes for half a century together.'[1] The work of a modern scholar confirms this last statement. According to Dr. Gilboy, at Westminster and Southwark the wages of the mason's labourer remained fixed at 2s. a day for more than fifty years after 1731, and at Oxford the pay of the builder's labourer stood at approximately 1s. 3d. a day from 1700 to 1770. Generally the wages of skilled workers were less rigid. But over the period 1700-1790 the daily rate of bricklayers, masons, and plumbers at Westminster and Southwark never fell below 2s. 6d. or rose above 3s. For seventy-eight years the plumber's daily wage stood at 3s., and for fifty-six years the plasterer's wage never varied from the same figure.[2] Nor was this rigidity peculiar to building. At Griff colliery in Warwickshire in 1729, the normal daily wage of 1s. 6d. was the same as had existed in 1701. At Barlow colliery in Derbyshire the same rate of 1s. 6d. was paid between 1763 and 1776.[3] And at Coalbrookdale the price paid to the chartermaster for delivery of a load of coal showed no change between 1768 and 1780. Ten shillings a week was the normal wage of a furnace keeper in the iron industry of north Lancashire up to 1755, and 10s. 6d. from this time to the end of the century. At the Horsehay ironworks in Shropshire between 1777 and 1781 the keeper of the furnace received 11s. a week, and in 1796, when the cost of living had risen substantially, his wage had risen only to 12s. 6d.[4]

In pursuit of his thesis, Adam Smith goes on to show that wages varied more from place to place than did the price of provisions. The goods bought by the workers were 'generally fully as cheap or cheaper in great towns than in remote parts of the country'. 'But', he continues, 'the wages of labour in a great town and its neighbourhood are frequently a fourth or a fifth part . . . higher than at a few miles distance.' Here, again, Dr. Gilboy's study provides illustrations. Wages of building workers were higher at Exeter, Oxford, Bristol, and Gloucester, than in the areas surrounding these places. The same was true of the wages of labourers in the North Riding and Lancashire, though (owing perhaps to the greater mobility of master craftsmen and journeymen in the north) the local differences in the earnings of skilled labour were less pronounced than in the

---

[1] *Op. cit.*, p. 66.
[2] E. W. Gilboy, *Wages in Eighteenth Century England*, App. II.
[3] T. S. Ashton and J. Sykes, *op. cit.*, p. 137.
[4] Horsehay MSS. In addition, at both periods, he had a bonus on output.

south-west. The textile industries offer examples of the same variation. In August, 1771, the weaver at, or near, Halifax, where the employer had his headquarters, received 11*d*. for his warp. His fellow, employed by the same firm but living 'in Lancashire and at a distance', had 10*d*. The difference may have represented the increased cost to the employer of putting out yarn and collecting warps at remote places. In periods of depression, when the price of the product fell, the regional difference in piece-rates widened— perhaps because the costs of distribution and transport now swallowed up a larger proportion of the receipts. In March, 1774, the pay of the Halifax weaver had fallen to 8*d*., that of the Lancashire weaver to 6*d*.[1]

The higher rates of pay offered in some areas may have reflected greater skill and closer division of labour. It was perhaps for this reason that, towards the end of the century, the miners of Cornwall obtained wages varying from 9*s*. to 15*s*. a week, against the 6*s*. to 10*s*. paid in Anglesey. (But conditions in the two areas were, in fact, hardly comparable. The Cornish miner had to toil in deep and dangerous places: in Anglesey the ore could be obtained largely from open-cast workings.) That relatively scarce skill obtained a higher reward than common labour could be illustrated by figures from a dozen separate trades. Adam Smith declared that the wages of a Newcastle pitman were generally double those of the labourer, and that the mason or bricklayer received from 'one half more to double' those of his unskilled assistant. (Surveys in other industries suggest that the differential was nearer to 50 than 100 per cent.) Difference of skill is one reason for the disparity in rates of pay between London and the provinces. The London craftsman was highly specialised: from the nature of things, the so-called skilled worker of the country districts was often a jack-of-all-trades. Adam Smith, however, offered a further explanation. Wages, he declared, varied with the constancy or inconstancy of employment. It was partly because of the periods of enforced idleness to which they were subject that the London tailors got the relatively high wage of 2*s*. 6*d*. a day; and the still higher degree of unemployment to which the coal-heavers were liable was adduced as the cause of their being able to make as much as 6*s*. or 10*s*. a day when at work. Adam Smith regarded such high rates of pay as compensation 'for those anxious and desponding moments which the thought of so

[1] J. Bischoff, *op. cit.*, p. 185.

precarious a situation must sometimes occasion'. When in 1771 the Nottingham stockingers asked for an increase in pay, they pointed to the higher earnings of the shoemakers and tailors. The reply of the masters was that in these occupations employment was seasonal, whereas the stockingers were able to work all the year round. But, ultimately, it was not considerations of distributive justice that determined the rates of wages. If workers subject to abnormal unemployment had not been offered higher pay they would have transferred their labour to other occupations. If their skill was such as to make this impossible, they would at least have seen to it that their sons did not enter the trade; and so, in the long run, a shortage of workers would have brought about an upward adjustment of earnings. If it had been simply a matter of equity there were large numbers of women, and scores of thousands of labourers in London, who might have established as convincing a claim to high wages as that of the building workers or the coal-heavers.

Differences of race, age, and sex operated through demand or supply, to produce marked differences in rates of wages. The Jews, most of whom were refugees, were in a weak position in the labour market: it was almost impossible for them to be apprenticed to a Christian, and, since they could not work on the Sabbath, most employers were unwilling to take them even as unskilled labourers. No wonder that many of them took to peddling, dealing in secondhand clothes, and uttering false money.[1] The immigrant Irishman was not subject to the same disabilities. But his religious beliefs, his pugnacity, and his habits of life made him uncongenial as a workmate to Englishmen. The strength and endurance of the Irish, however, enabled them to get employment—usually at relatively low rates of pay—as harvesters, market porters, bricklayers' labourers, coal-heavers, and dock workers. Like the Jews, they and their womenfolk were often hawkers and costermongers, and some of them took to anti-social activities. In the later decades of the century, when over-population at home drove large numbers to England, they helped to meet the needs for labour of the rapidly expanding cotton industry. Few, however, entered the coal or iron industries except as unskilled labourers. Social and national feeling restricted their field of work and kept their wages at a relatively low level.

Similar barriers stood in the way of female workers. Women formed the greater part of the labour force in the textile industries,

---

[1] M. D. George, *op. cit.*, pp. 125-32.

but their work was restricted to the earlier processes of wool picking, cleaning, and spinning. Spinning was hardly a craft: it was, for many, a spare-time occupation, like knitting today, and those who looked on it as a sole means of maintaining themselves had to live at a low standard of comfort. When the mule was introduced into the cotton industry it was supposed to require the skill and strength of a man: its output was so much greater than that of the spinning wheel as to force down the rates paid to women spinners on the finer counts. As their menfolk forsook the loom for the mule, however, many women began to weave; and the opening up of this new field of activity did something to raise their status and earnings. In the silk industry, women had taken to weaving at an earlier period, but in London the greater part of the women and girls in the trade were in the ill-paid branches of winding and doubling. During the period of prosperity that came with the Seven Years' War more of them seem to have worked on the loom; but in 1769 the men sought to exclude them from the weaving of handkerchiefs—though it was agreed that if war broke out again the masters should be at liberty to employ them in any branch of the trade they might think fit.[1] Again, in the hosiery industry of London, though not in that of the Midlands, they seem to have been excluded from work on the knitting frames. This meant that the supply of women available for the smaller trades of London was excessive in relation to the demand for their services. Many of them were employed by their husbands or other outworkers: their story makes depressing reading.

In the heavy industries few women found opportunities of employment. It is rare to find a reference to any woman employed underground in the coal pits of Northumberland and Durham, at least in the later decades of the century, though girls were employed at the pit head in the arduous task of picking out stone from the coal. In the smaller mines of Lancashire, Yorkshire, and the Midlands, however, they frequently worked as carriers in the underground ways. At ironworks they were employed in such humble tasks as sifting the ore at the furnace, but there was no scope for their labour in the main processes.

The attempt of adult, male workers to limit the field of competition for employment had a similar effect on the earnings of the young. It was held in the law courts that the apprenticeship clauses of the Act of 1563 did not apply to trades that had come into being

[1] M. D. George, *op. cit.*, p. 182.

since that date. 'The manufacturers of Manchester, Birmingham, and Wolverhampton', wrote Adam Smith, 'are many of them, upon this account, not within the statute.'[1] In most of the corporate towns, however, it was usual to insist that every boy should serve for seven years under indentures before setting up in a skilled trade. In other places, too, apprenticeship continued, partly because the fees paid on indenture were a consideration to small-scale employers, but mainly because the workers found it a useful means of keeping down the number of journeymen, and so of maintaining the level of wages. The attitude of Parliament varied. Generally it was opposed to restrictions on the freedom of employers to hire whatever kind of labour they wished. But after 1710, when a revenue duty was imposed on articles of indenture, the state had a vested interest in the maintenance of apprenticeship.[2] In 1725 an act was passed restricting the making of broadcloth in the West Riding to those who had served as apprentices for seven years. But in 1733 this measure was repealed; and in 1751 a committee of the House of Commons reported strongly against compulsory apprenticeship.[3] It was clear that this was inappropriate to a system of industry in which technical change was an almost daily occurrence. Adam Smith inveighed against the custom on grounds of natural justice: 'The patrimony of a poor man lies in the strength and dexterity of his hands; and to hinder him from employing this strength and dexterity in what manner he thinks proper without injury to his neighbour is a plain violation of this most sacred property.' He asserted that long apprenticeship tended to idleness; that youths would be more diligent if, from the start, they were employed as journeymen; and that freer competition among workers would lead to a reduction of costs.[4]

Enough has been said to indicate the causes of the variations of wages between social groups. Something may now be said of variations over time. The inelasticity of rates of pay in the building trades was not representative of wages in industry as a whole. Generally, as production extended earnings rose. But there were periods of temporary expansion or contraction that led to wide fluctuations of wages and, in the textile industries in particular, these were such as must have resulted in still greater fluctuations of standards of living. Once again we may turn to *The Wealth of Nations*: 'The

[1] *The Wealth of Nations*, vol. I., p. 109. [2] E. Lipson, *op. cit.*, p. 290.
[3] *Ibid.*, pp. 289-90. [4] *The Wealth of Nations*, vol. I, pp. 110-12.

variations in the price of labour', we are told, 'not only do not correspond either in place or time with those in the price of provisions, but they are frequently quite opposite'. The effects of good and bad harvests on the demand for labour in agriculture have already been noticed: a bad harvest might increase the income of farmers and dealers, but it probably did not greatly increase their expenditures; and since a high price of food left the rest of the people with less to spend on other things, it decreased the demand for workers in many occupations. In the dearth of 1728 the wages of spinners in Lancashire were reduced by a third: in the good harvest of 1730 they were advanced by a quarter.[1] The famine year of 1741 saw a fall in the piece rate from 7d. to 5d. and a sharp rise in unemployment. In February of this year the *Gentleman's Magazine* reported that 'at Spittle-fields and in diverse parts of the Isle the manufacturers are starving for want of work'; and in the dearths of 1756-7, 1766-7, and 1772-4 there was the same unhappy concurrence of rising costs of living, falling wages, and unemployment. It was on the poorest that the burden of a failure of the harvest bore most heavily. According to one witness in 1774, the wages of the Yorkshire woolcomber were reduced by $12\frac{1}{2}$ per cent., those of the weaver by $16\frac{2}{3}$ per cent., and those of the spinner by 27 or 28 per cent.[2] Since the ratio of weavers to combers was three to one, and that of spinners to weavers at least as great, the classes most heavily hit far exceeded in numbers those who escaped with a relatively small curtailment of income. It would be easy to continue the story to the end of the century, but enough has been said to indicate the perverse elasticity that characterised wages in the textile industries.

Other influences from the side of demand often caused short-term changes in wages. The death of a monarch or a member of the royal family might lead to a sudden increase in the earnings of tailors and dressmakers. The period of civic improvements in London was marked by a rise in the daily earnings of the paviours between 1763 and 1771.[3] The outbreak of a war which cut off foreign markets sometimes led to a reduction of wages in the export industries. On the other hand, it might result in a rise of earnings in manufactures like that of silk, which were normally exposed to foreign competition, but which now had a monopoly of the home market. The American war, with its demands on the shipyards, led to a sympathetic

[1] *Supra*, pp. 60-1. [2] J. Bischoff, *op. cit.*, vol. I, p. 186.
[3] E. W. Gilboy, *loc. cit.*, p. 257.

rise in the wages of carpenters in the building trades of London. But nowhere was the effect of war on wages more clearly exemplified than in the case of the seamen. Large numbers left the merchant vessels to join the fighting ships or engage in privateering; and in spite of an influx of foreign seamen, rates of pay were far in excess of those offered in times of peace. During the war of 1776-83 wages increased from £1 10s. to £3 5s. a month. In that of 1793-1802 the collier vessels were manned by a motley collection of indentured apprentices, foreigners, men rejected by recruiting officers because of infirmity, and (in the autumn and winter) seamen from the Greenland fisheries. The last are said to have been the 'ringleaders of all disturbances for raising wages', and their efforts were highly successful. The crews were paid by the voyage—a round trip from Newcastle to London and back that might take a month or six weeks. In 1792 (a year of brisk trade) the usual rate was £2 10s. a voyage. Immediately on the outbreak of war, in February, 1793, it rose to £3 10s.; by April it was £7 17s. 6d., and by the end of the year £8 1s. 6d. In July of 1794 it fell to £5 5s. (rates were always lower in summer than in winter), but in November it stood at £8. The pay continued to fluctuate, but the trend was upward: in 1796 as much as £10 a voyage was paid, and in April, 1800, the men were receiving 'the enormous rate' of eleven guineas a voyage. The cost of living was rising, but it can hardly be denied that, in this calling, labour did well out of the war.[1]

IV

Something has already been said of the organisations of the employers. These were not without influence on national policy as to labour. Industrial legislation was directed not only to regulating wages and hours, as in the case of the tailors and silk weavers, but to suppressing embezzlement and preventing skilled workers from going overseas.[2] But, even without such aid, the manufacturers were in a strong position in bargaining with their employees: associations set up for other purposes could be used to control labour. There was no matter on which opinion was less divided than on the wickedness of poaching on another man's supply of labour. As has already been said, in many industries it was usual to hire men by the year, and to impose penalties for infringement of the contract. In the

[1] *Report on the State of the Coal Trade* (1800), App., pp. 553, 563.
[2] For the emigration of artisans see especially the Home Office papers for 1749-59.

domestic trades a wage-earner who left an employer before completing work given out to him was liable to prosecution, and if a man owed money to his master anyone who wished to engage him was expected to take responsibility for the debt. But even if there were no such complication it was considered improper to hire a workman without the consent of the previous employer. Surviving business records contain many letters in which charges are made of seducing labour, and nearly as many in which the accused declares himself wounded at the suggestion that he should have been thought capable of such unprofessional conduct.

The workers, however, were by no means unable to defend their interests. Most of them, like most employers, thought in terms of customary standards of life. Hence it was in times of dearth that their corporate sense was most clearly, and violently, manifested. Generally it was not the employers, but the bakers, millers, dealers, and, above all, exporters of corn who were held responsible for the distress. Following the harvest failure of 1709 the keelmen of the Tyne took to rioting. When the price of food rose sharply in 1727 the tin-miners of Cornwall plundered granaries at Falmouth, and the coal-miners of Somerset broke down the turnpikes on the road to Bristol. Ten years later the Cornish tinners assembled again at Falmouth to prevent the exportation of corn, and in the following season there was rioting at Tiverton. The famine of 1739-40 led to a 'rebellion' in Northumberland and Durham in which women seem to have taken a leading part: ships were boarded, warehouses broken open, and the Guildhall at Newcastle was reduced to ruins. At the same time attacks on corn dealers were reported from North and South Wales. The years 1748 and 1753 saw similar happenings in several parts of the country; and in 1756-7 there was hardly a county from which no report reached the Home Office of the pulling down of corn mills and Quaker meeting-houses, or the rough handling of bakers and grain dealers. In spite of drastic penalties the same thing occurred in each of the later dearths of the century: in 1762, 1765-7, 1774, 1783, 1789, 1795, and 1800.[1]

It is clear that such disturbances were not industrial in character. Others that may be so termed arose from circumstances over which individual employers had little control. Throughout the century, and especially in times of depression, animosity against the immigrant Irish, who worked for low rates of pay, led to tumults.

[1] T. S. Ashton and J. Sykes, *op. cit.*, ch. VIII.

Anti-Irish sentiment probably played a larger part in the fight between the sailors and the coal-heavers in 1768, when conditions were so serious that 'the work of every journeyman gunsmith out of the Tower, done or undone, is called in for fear it should fall into desperate hands'.[1] In 1774 there were pitched battles between English and Irish haymakers around Kingsbury, Hendon, and Edgware;[2] and opposition to Irish labour played a part in the Gordon riots of 1780. In other cases troubles arose from protests against the import of goods that competed with those made by English workers: in 1764 several thousand London weavers demonstrated against the clandestine importation of French silks.[3]

When regular unions existed, it was natural that protests made against high prices of food or 'unfair' labour should sometimes have been accompanied by demands for higher earnings. In the corn riots of 1740 the northern pitmen declared their intention of enforcing an increase of wages,[4] and, in the same period of dearth, 'a battallion of Guards and a Troop of Horse march'd to Woolwich to quiet the Workmen in that Yard, who mutiny'd about their Pay and refus'd to work'.[5] As time went on such demands became more frequent: in 1778, for example, a petition to Parliament by the frame-work knitters of Nottingham calling attention to the high cost of provisions, asked for a statutory increase of wages and an abatement of frame-rents.[6] When rioting was accompanied by destruction of machinery animosity against employers is obvious. In some cases, as in the pulling down of gig-mills by the shearmen, and the destruction of Arkwright's mill at Birkacre in 1779, fear of labour-saving devices was clearly operative. But the same cannot be said of the action of west-country weavers who destroyed tenter-frames, or of the colliers who frequently smashed the pit gear, and sometimes even set the mines on fire: they must have realised that their action would result in unemployment, but their immediate concern was to assert their strength and inflict loss on stubborn employers. There seems to have been little or no social theory in the minds of the rioters and very little class-consciousness in the Marxist sense of the term. Some of the early trade societies, such as those of the smallware weavers of 1747, were associations of journeymen and small masters: it would have been difficult to enlist their members

[1] *Annual Register*, May and June, 1768. [2] M. D. George, *op. cit.*, p. 357.
[3] *Annual Register*, April 9th, 1764. [4] T. S. Ashton and J. Sykes, *op. cit.*, p. 118.
[5] *Annual Register*, November 5th, 1739.
[6] J. D. Chambers, *Nottinghamshire in the Eighteenth Century*, p. 40.

in any movement for the overthrow of the existing order. Differences between one grade of worker and another were often as acute as those between wage-earners and employers. In 1749 trouble arose at Tiverton about the importation of Irish worsted yarn. The woolcombers were strongly opposed to this, but the weavers, whose livelihood depended on the Irish material, took sides with their employers; and the two groups of workers engaged in fierce battles.[1] In 1767 rusty swords and pistols were displayed, if not used, in a fight between the 'narrow weavers' and 'engine weavers' of Spitalfields; and objection to the employment of women in some branches of the silk trade seems to have played a part in the disturbances that broke out two years later.[2] In 1765 the Tiverton weavers rioted on behalf of a would-be employer, Charles Baring, who was willing to set up in the town but whose application to be made a freeman had been rejected by the mayor and aldermen.[3]

It would be wrong to lay stress on the part played by violence in the efforts of the workers to maintain and improve their standards of employment. The scattered women spinners, the labourers, and many of the weavers were too poor to establish anything in the nature of trade unions. But among the skilled artisans friendly societies and box-clubs flourished. Wool-combers, dyers, tailors, cabinetmakers, printers, cutlers, grinders, shipbuilders, keelmen, sailors, and many others had their separate organisations in the first half of the century. Often the chief function was to insure against the more serious effects of sickness and unemployment; but most of the societies had also industrial objects. Their aim was to keep down the number of those in the particular trade and so to preserve local standards of life. Hence the maintenance of strict rules of apprenticeship played a large part in their policies. The employment of those who did not contribute to the 'box' was resisted; and the financially stronger among the unions sought to prevent local unemployment from forcing down wages, by the grant of travelling benefit. Where, as in the case of the keelmen of Newcastle, work was concentrated in a limited area, and where, as in their case also, there was an assured market, it was possible by joint action to raise wages well above the level in other occupations. And even where no such conditions existed, a group controlling an essential process could

---

[1] M. Dunsford, *op. cit.*, 230-2.
[2] *Annual Register*, November 30th, 1767; M. D. George, *op. cit.*, p. 182.
[3] M. Dunsford, *op. cit.*, pp. 245-52.

sometimes improve its position at the expense not only of the employers, but of other workers in the industry. 'Half a dozen woolcombers, perhaps, are necessary to keep a thousand spinners and weavers at work', wrote Adam Smith. 'By combining not to take apprentices they can not only engross the employment, but reduce the whole manufacture into a sort of slavery to themselves, and raise the price of their labour much above what is due to the nature of their work.'[1] The wool-combers lived close to one another in towns, and, though the worsted industry was carried on in several regions, the number of centres of employment was limited. Local surpluses of labour could be dispersed by issuing to unemployed men 'blanks' or certificates, which gave them a claim to support by every woolcombers' society throughout England. The millwrights also made use of tramping benefits. According to James Watt and other employers, they were more than adequately paid for their skill.

Less favoured groups of workers sought to improve their standards by bargaining and local strikes. Among those paid by time, and not by the piece, frequent attempts were made to reduce the length of the working day: a shorter day meant an increase of hourly rates. In tailoring and the building trades, in particular, the question whether men should work for twelve hours or for longer was the occasion of several strikes. Many unions, including those of the coal miners, had rules as to the amount of work to be done each day or each week. Restrictions on output were condemned by most observers as injuring not only the public, but those who conformed to them. Where the demand for the service was inelastic, however, as in some of the crafts in the building industry, and in the unloading of coal on the Thames, there was an obvious sectional benefit. Both contemporary and later writers often attributed the decline of an industry in a locality such as Spitalfields or Norwich to restrictive practices among the workers. In some cases, however, it is probable that the sequence ran in the opposite direction: the decline of the industry, for whatever reason, led to local unemployment, and the restriction was an attempt—ineffective though it might be—to stem the course of this.

Some of the unions, the weaker especially, sought to enlist support from Parliament. Again and again, appeals were made to enforce the clauses relating to wages in the Statute of Artificers of 1563. But when, as in the case of the tailors in 1721, the legislature agreed

---

[1] *The Wealth of Nations*, vol. I, p. 114.

to the determination of wages and hours by the magistrates, the permission was accompanied by a prohibition of trade unions. In 1726 combination was forbidden in the woollen trade, and in 1749 in the silk, linen, cotton, fustian, iron, leather, and several other industries. These measures alone are evidence that, even at this relatively early stage of industrial development, combinations of wage-earners were widespread. At the end of the century, in 1799, Parliament passed a measure which prohibited combination of labour universally. But in view of what had gone before, this was of less account than has usually been thought. Employers who found their progress impeded by trade unions could bring actions for conspiracy under the common law, and they could also prosecute individuals for laying down their tools before completing their allotted tasks. In fact, it would seem, the law did little to check the growth of workers' combinations. The rapid expansion of the cotton industry in the 'nineties was accompanied by the growth of powerful unions of skilled spinners. And, at the same time, an efflorescence of friendly societies, protected and stimulated by Rose's Act of 1793, gave cover to many organisations formed primarily for industrial purposes.

How far concerted action raised wages, prevented their fall, or in other ways benefited the workers it is impossible to say. That it did little for the poorest of all is certain. One may hazard the opinion that such improvement as occurred in the lot of most English wage-earners was the result not of institutions, but of economic forces discussed in earlier chapters.

V

Adam Smith held that 'it is not the actual greatness of national wealth but its continual increase, which occasions a rise in the wages of labour'. This was a matter on which he laid great stress. 'It is not, accordingly', he adds, 'in the richest countries, but in the most thriving, or in those which are growing rich the fastest, that the wages of labour are highest.' In a later passage he returns to the theme: 'It is in the progressive state, while the society is advancing to the further acquisition, rather than when it has acquired its full complement of riches that the condition of the labouring poor, of the great body of the people, seems to be the happiest and the most comfortable. It is hard in the stationary, and miserable in the declining state.'[1]

[1] *Loc. cit.*, vol. I, p. 72.

In few parts of England was the economy stationary or declining, but the pace of advance varied greatly from one region to another. It was far more rapid in the north than in the south—as was pointed out by observers as early as the 'thirties. Professor Tawney has remarked that Lancashire in the eighteenth century had most of the features of a colony. But colonies are of varying types: the one to which Lancashire conformed was that which a distinguished Canadian economist has described as the pioneer economy.[1] Like Canada today, Lancashire was relatively under-populated; the rate of investment (and hence of saving) was high; and the demand for labour was such as to lead to a progressive increase in rates of wages. At the beginning of the century the normal rate of unskilled workers in Lancashire seems to have been about 8d. a day. It stood in sharp contrast to the 14d. paid in Oxford and the 20d. in London. By the middle of the century the Lancashire rate had risen to 12d.; the Oxford rate still stood at 14d.; and the London rate was 24d. By the late 'eighties Lancashire wages, at 20d., were well above the Oxford level of 16d., and approached the 24d. that was still the normal rate in London. At the end of the century Lancashire almost certainly stood in the forefront, for by 1793 the rate had touched 24d., and it is highly unlikely that no further increases occurred.[2]

If it were possible to obtain series of wages for artisans it is probable that even larger advances would be found in the areas where industrial development was most rapid. Birmingham was expanding more rapidly than Sheffield, and wages there were generally higher. Adam Smith attributed the difference to the fact that many of the Birmingham trades were of recent creation: 'When a projector attempts to establish a new manufacture, he must at first entice his workmen from other employments by higher wages . . . and a considerable time must pass away before he can reduce them to the common level.'[3] But rapid as was the course of innovation in Birmingham, it hardly kept pace with that in the industrial north: the wages of men at the Soho works of Boulton and Watt were considerably less than those paid in Manchester, where, in 1792, it was said (though perhaps with some exaggeration), 57s. or 58s. a week was the minimum paid to skilled workers.[4]

[1] W. A. Mackintosh, 'Aspects of a Pioneer Economy', *Canadian Journal of Economics and Political Science*, vol. II, no. 1 (1936).
[2] E. W. Gilboy, *op. cit.*, App. II.
[3] *The Wealth of Nations*, vol. I, p. 103.
[4] Eric Roll, *An Early Experiment in Industrial Organisation*, p. 190.

In the last two decades of the century, not merely individual regions, or industries, but the economy as a whole felt the quickening influence of improvements in transport and the introduction of cheap iron and steam power. The innovations of Arkwright, Watt, and Cort called into being new grades of machine makers, engine erectors, and skilled operatives. Social changes increased the mobility of labour. Men who had served in the wars had become accustomed to being on the march. And the appearance of groups of canal workers in a village must have encouraged the inhabitants to seek opportunities of work outside their own areas. The spread of education, by Sunday schools and other bodies, the sight of advertisements in the newspapers that were springing up in the provinces, the payment of tramping benefits by the trade unions, the modification of poor-law regulations—these and other influences widened the market for labour. The effect was to reduce local variations of wages and to assimilate the rates paid in the country to those of the towns. England, and not merely Lancashire, was entering 'the progressive state'.

More information is available about wages in these decades than in any earlier period; and it would appear that for the skilled worker, in agriculture and manufacture alike, the trend was upward.[1] It would be rash, however, to use the material assembled by several scholars to make generalisations as to changes in the standard of living. We do not know whether allowances in kind were increasing or declining; whether the imposition of fines was growing or diminishing; whether indebtedness to employers was becoming more, or less, common; whether periods of unemployment were lengthening or shortening; whether opportunities for overtime work were more numerous or fewer. Earnings varied with the age of the worker, but few wages books record whether the person concerned was a child, a youth, or an adult. Even if there were clear information as to net weekly payments to men, there would still remain the unanswerable question of the number of earners and dependants in the family. Still more formidable difficulties confront attempts to measure real wages.[2] There is no index of the cost of living, and if one were devised it could not take account of local variations of diet, which

---

[1] See the series of scholarly papers contributed to the *Journal of the Royal Statistical Society*, from 1898 onward, by Sir Arthur Bowley, and especially those in vols. LXI, LXII, and LXIV.
[2] Such as that made by Rufus S. Tucker, 'Real Wages of Artisans in London, 1729-1935', *Journal of American Statistical Association*, vol. XXI, p. 73 f.

were still marked. To divide a series of wage-rates by a series of prices paid, under long-term contracts, by schools and government departments, is to obtain a result of little meaning. The few observations that follow must be regarded, therefore, as subject to correction by researches now in progress.

It would seem that, as factory production developed, payment of wages became more regular; and under-employment also diminished. On the other hand, the 'nineties were marked by disturbances of industry arising from war, inflation, and, above all, variations in the yield of the crops: any statement of how much a family wage would buy in, say, 1792 or 1794 would be false if applied to 1795 or 1800. Throughout the two decades, the prices of agricultural products were rising more steeply than those of manufactured goods: whatever may have been true of the movement of prices of imports and exports, within England itself the terms of trade were moving against the manufacturing classes. It would seem that the difference in pay between skilled and unskilled increased. The notion, widely prevalent, that machine production reduced the need for skill, receives no support from the evidence available: the career was open to the talents, and manual dexterity, no less than enterprise, received its reward. The skilled man, with a rising income, had a margin to spend on the products of manufacture, and the range of his choice in expenditure widened. If he were not becoming better off it would be difficult to explain the increased consumption of tea, sugar, beer, and the cheaper forms of textile goods that occurred at this time, or the rise of nonconformist chapels, friendly societies, and trade unions.

But what of the rest of the wage earners? Thoughtless writers[1] have compared the semi-skilled operatives in the new factories with the small farmers and craftsmen of an earlier generation. If comparison is to be made at all it must be with the squatters of the countryside, and the paupers of the towns, from whose ragged ranks the factory workers were largely drawn. The greater part of them consisted of women, young people, and children. Their lot was unenviable, but at least it was no worse than that of their forebears: there is no evidence that many of them starved. Outside the factories were skilled weavers of fine fabrics, metal workers, and others who were able to maintain their old standards. But there were also domestic spinners, weavers of plain cloths, and workers in a variety of small trades, whose services were no longer much in demand.

[1] Following the romantic Gaskell and the doctrinaire Engels.

And the growth of population, together with Irish immigration, had swelled the numbers of common labourers, as well as of the poor who had never engaged in sustained work. The greater part of the expenditure of these was on food. The trend of food prices was upward, and, though the same was true of the labourer's wage, it was not so of the earnings of many thousands of domestic workers. When famine appeared in 1795-6 and 1800-1 every penny was needed to keep body and soul together. Some died of sheer hunger, and but for a lenient administration of the poor law there must have been social disaster. The divergence of experience between skilled and less skilled explains how honest observers could differ as to whether things were getting better or worse for labour at this time. And it is because historians tend to look at this or that section of the workers exclusively that the controversy continues.[1]

If among the well-to-do there were men, like Temple, who despised the lower orders, there were not lacking others, like Coram, Hanway, and Fielding, who showed sense and humanity. The generations that are the subject of this volume passed on to their successors a legacy of poverty hardly less great than that they had themselves inherited. But they left, also, some tradition of philanthropy and public spirit. The nineteenth century was to see a remoulding of institutions and a change of attitude to the problem of poverty. But it was less by reason of these than of economic and demographic change, that, in this maligned period, 'the progressive state' paid its dividend, not only to capitalists and skilled workers, but also to the labouring poor.

[1] This theme is developed in a paper by the author, 'The Standard of Life of the Workers in England, 1790-1830', *Journal of Economic History*, Supplement IX (1949), pp. 19-38.

APPENDIX

Statistical Tables

I	Prices of Wheat, 1704-1800	239
II	Output of Hops, 1712-1800	240
III	Output of Malt, 1702-1800	241
IV	Output of Strong Beer, 1700-1800	242
V	Output of British Spirits, 1700-1800	243
VI	Output of Starch, 1713-1800	244
VII	Cattle and Sheep brought for sale to Smithfield, 1732-1794	245
VIII	Tallow Candles Charged with Duty, 1711-1800	246
IX	Soap Brought to Duty, 1713-1800	247
X	Printed Goods Charged with Duty, 1713-1800	248
XI	Production of Broadcloths in the West Riding, 1727-1800	249
XII	Production of Narrow Cloths in the West Riding, 1738-1800	250
XIII	The Average Price of Bank Stock 1700-1800, of 3% Annuities 1726-57 and of 3% Consolidated Stock 1758-1800	251
XIV	Total English Exports and Total Imports, 1700-1800	252
XV	Rate of Sterling Exchange on Hamburg, 1700-1800	253
XVI	Bankruptcies, 1732-1800	254

Table I

PRICES OF WHEAT PER QUARTER AT MICHAELMAS, 1704-1800

Year	s.	Year	s.	Year	s.
1704	24	1737	31	1769	38
05	23	38	25		
06	20	39	34	1770	40
07	24			71	49
08	39	1740	45	72	58
09	80	41	34	73	60
		42	24	74	64
1710	60	43	21	75	46
11	47	44	21	76	48
12	29	45	25	77	46
13	48	46	28	78	44
14	32	47	29	79	36
15	36	48	36		
16	42	49	33	1780	34
17	40			81	46
18	24			82	61
19	25	1750	31	83	64
		51	37	84	60
1720	34	52	40	85	54
21	28	53	35	86	29
22	30	54	28	87	48
23	32	55	29	88	48
24	32	56	48	89	62
25	47	57	52		
26	38	58	46	1790	58
27	42	59	31	91	48
28	54			92	58
29	47	1760	30	93	52
		61	27	94	48
1730	34	62	35	95	92
31	24	63	37	96	65
32	20	64	52	97	54
33	21	65	45	98	54
34	33	66	50	99	93
35	36	67	60		
36	34	68	45	1800	128

The prices are given to the nearest shilling. Those from 1704 to 1794 are for Cambridge and are taken from J. E. Thorold Rogers, *A History of Agriculture and Prices in England*, Vols. VI and VII, Pt. I. From 1795 the figures are for London and have been taken from Thomas Tooke, *A History of Prices*, Vol. II, p. 387.

Table II

OUTPUT OF HOPS, 1712-1800
(in thousands of lb.)

Year ending June 24th		Year ending June 24th		Year ending June 24th	
1712	12,005	1742	17,984	1772	7,968
13	7,263	43	10,916	73	24,618
14	5,531	44	15,330	74	11,013
15	3,466	45	11,211	75	33,320
16	10,792	46	8,339	76	10,004
17	4,882	47	22,063	77	30,196
18	13,109	48	15,160	78	10,501
19	3,601	49	20,951	79	38,416
1720	21,665	1750	8,713	1780	12,738
21	9,159	51	17,323	81	27,928
22	14,764	52	17,796	82	25,861
23	11,938	53	19,725	83	3,112
24	7,266	54	21,875	84	15,802
25	14,688	55	26,962	85	19,736
26	1,566	56	19,820	86	23,519
27	20,392	57	11,523	87	19,996
28	16,659	58	16,749	88	8,814
29	9,955	59	17,506	89	29,978
1730	11,626	1760	10,107		
31	10,636	61	28,437	1790	21,177
32	5,583	62	19,606	91	22,308
33	8,408	63	18,993	92	18,868
34	16,868	64	21,204	93	33,950
35	8,987	65	4,139	94	5,000
36	10,316	66	17,752	95	42,529
37	11,134	67	27,988	96	17,180
38	13,619	68	6,245	97	15,740
39	20,962	69	27,440	98	32,881
				99	11,775
1740	16,977	1770	3,891		
41	9,091	71	24,299	1800	15,293

Customs Library, *Excise Revenue Accounts, 1662-1867*.

Table III
OUTPUT OF MALT, 1702-1800
(in millions of bushels)

Year ending June 24th		Year ending June 24th		Year ending June 24th	
1702	12·6	1735	26·3	1768	28·0
03	27·6	36	24·4	69	27·4
04	20·4	37	25·3		
05	28·0	38	27·0	1770	25·2
06	23·8	39	27·6	71	22·6
07	25·8			72	28·4
08	23·9	1740	22·8	73	22·1
09	20·9	41	20·8	74	24·7
		42	26·7	75	25·8
1710	20·3	43	27·1	76	27·2
11	23·0	44	32·8	77	26·6
12	23·0	45	25·7	78	27·1
13	25·9	46	24·7	79	27·1
14	20·7	47	25·7		
15	25·2	48	27·3	1780	31·8
16	27·5	49	25·8	81	27·6
17	29·8			82	28·0
18	27·7			83	17·2
19	29·1	1750	30·2	84	26·6
		51	27·8	85	27·1
		52	25·0	86	22·8
1720	26·4	53	26·0	87	27·3
21	29·5	54	28·2	88	26·9
22	34·0	55	28·8	89	24·3
23	31·6	56	24·9		
24	25·0	57	18·2		
25	28·1	58	25·8	1790	22·7
26	27·9	59	29·0	91	27·9
27	26·2			92	28·7
28	21·6			93	24·5
29	23·8	1760	28·7	94	25·6
		61	29·8	95	24·7
		62	26·8	96	28·1
1730	29·3	63	20·2	97	30·9
31	26·6	64	27·2	98	27·0
32	27·8	65	26·4	99	31·8
33	30·7	66	21·5		
34	27·9	67	22·6	1800	14·5

British Parliamentary Papers, 1835, XXXI, p. 346.

Table IV

OUTPUT OF STRONG BEER, 1700-1800
(in millions of barrels)

Year ending June 24th		Year ending June 24th		Year ending July 5th	
1700	3·15	1735	3·52	1767	3·62
01	3·36	36	3·55	68	3·73
02	3·72	37	3·61	69	3·79
03	3·65	38	3·53		
04	3·76	39	3·60	1770	3·78
05	3·81			71	3·77
06	3·63	1740	3·52	72	3·79
07	3·65	41	3·30	73	3·80
08	3·57	42	3·46	74	3·60
09	3·36	43	3·47	75	3·86
		44	3·60	76	3·96
1710	3·22	45	3·48	77	4·11
11	3·15	46	3·42	78	4·13
12	3·14	47	3·58	79	4·20
13	3·30	48	3·62		
14	3·42	49	3·71	1780	4·36
15	3·43			81	4·34
16	3·49	1750	3·74	82	4·52
17	3·60	51	3·77	83	3·91
18	3·72	52	3·77	84	4·34
19	3·80			85	4·33
				86	4·15
1720	3·79	Year ending July 5th		87	4·43
21	3·94	1753	3·76	88	4·30
22	3·79	54	3·73	89	4·44
23	3·85	55	3·69		
24	3·87	56	3·78	1790	4·83
25	3·80	57	3·50	91	4·75
26	3·64	58	3·72	92	5·08
27	3·70	59	3·91	93	5·17
28	3·44			94	5·01
29	3·31	1760	4·14	95	5·04
		61	4·06	96	5·50
1730	3·53	62	3·80	97	5·84
31	3·64	63	3·76	98	5·78
32	3·70	64	3·82	99	5·77
33	3·69	65	3·69		
34	3·68	66	3·68	1800	4·82

Customs Library, *Excise Revenue Accounts, 1662-1827.*

Table V

OUTPUT OF BRITISH SPIRITS, 1700-1800
(million gallons)

Year ending June 24th		Year ending June 24th		Year ending June 24th		Year ending July 5th	
1700	1·23	1726	3·98	1751	7·05	1775	2·51
01	1·27	27	4·61	52	4·48	76	2·52
02	1·07	28	4·79			77	2·51
03	1·15	29	4·73	Year ending		78	2·96
04	1·38			July 5th		79	2·64
05	1·44	1730	3·78	1753	4·87		
06	1·67	31	4·33	54	5·05	1780	2·72
07	1·97	32	4·37	55	4·65	81	2·46
08	1·72	33	4·82	56	4·68	82	2·18
09	1·74	34	6·07	57	3·71	83	1·37
		35	6·44	58	1·85	84	1·50
1710	2·20	36	6·12	59	1·82	85	3·14
11	2·23	37	4·25			86	4·31
12	2·07	38	5·44	1760	2·32	87	3·22
13	2·05	39	5·76	61	3·18	88	2·87
14	1·95			62	2·30	89	4·04
15	2·27			63	2·27		
16	2·38	1740	6·65	64	2·22	1790	4·32
17	2·60	41	7·44	65	2·23	91	4·55
18	2·42	42	7·95	66	2·43	92	5·06
19	2·46	43	8·20	67	2·04	93	4·73
		44	6·63	68	2·21	94	5·12
		45	7·20	69	2·55	95	5·15
1720	2·48	46	6·87			96	0·87
21	2·79	47	7·31	1770	2·57	97	3·18
22	3·38	48	7·08	71	2·50	98	3·97
23	3·70	49	6·67	72	2·58	99	4·51
24	3·56			73	2·22		
25	3·93	1750	6·61	74	2·01	1800	4·84

Customs Library, *Excise Revenue Accounts, 1662-1827*.

Table VI

OUTPUT OF STARCH, 1713-1800
(in millions of lb.)

Year ending June 24th		Year ending June 24th		Year ending July 5th	
1713	2·69	1743	1·60	1771	2·80
14	2·46	44	1·59	72	3·52
15	2·68	45	1·53	73	3·68
16	2·53	46	1·48	74	5·37
17	2·86	47	1·68	75	4·94
18	2·75	48	1·91	76	5·55
19	3·11	49	2·02	77	4·87
				78	4·65
1720	2·86	1750	2·33	79	6·63
21	2·62	51	2·25		
22	2·72	52	2·38	1780	6·44
23	2·54			81	4·87
24	2·48	Year ending July 5th		82	6·94
25	2·35	1753	2·59	83	5·38
26	2·03	54	2·90	84	6·71
27	2·43	55	3·33	85	6·27
28	1·91	56	3·24	86	6·61
29	1·69	57	2·23	87	5·95
		58	3·40	88	6·07
1730	1·82	59	4·14	89	6·40
31	1·76				
32	1·97	1760	3·79	1790	6·97
33	1·88	61	3·45	91	7·78
34	1·66	62	3·66	92	8·45
35	1·36	63	3·62	93	7·28
36	1·43	64	4·00	94	7·92
37	1·54	65	3·56	95	8·02
38	1·40	66	3·78	96	1·91
39	1·37	67	3·57	97	3·14
		68	3·64	98	7·31
1740	1·09	69	4·12	99	5·88
41	0·94				
42	1·55	1770	3·90	1800	2·88

Customs Library, *Excise Revenue Accounts, 1662-1827*.

Table VII
CATTLE AND SHEEP BROUGHT FOR SALE TO SMITHFIELD, 1732-1794

	Cattle	Sheep		Cattle	Sheep
1732	76,210	514,700	1764	75,168	556,360
33	80,169	555,050	65	81,630	537,000
34	78,810	566,910	66	75,534	514,790
35	83,894	590,979	67	77,324	574,050
36	87,606	587,420	68	79,660	626,170
37	89,862	607,330	69	82,131	642,919
38	87,010	589,470			
39	86,787	568,980	1770	86,890	649,090
			71	93,573	631,860
1740	84,810	501,020	72	89,503	609,540
41	77,714	536,180	73	90,133	609,740
42	79,601	503,260	74	90,419	585,290
43	76,475	468,120	75	93,581	623,950
44	76,648	490,620	76	98,372	671,700
45	74,188	563,990	77	93,714	714,870
46	71,582	620,790	78	97,360	658,540
47	71,150	621,780	79	97,352	676,540
48	67,681	610,060			
49	72,706	624,220	1780	102,383	706,850
			81	102,543	743,330
1750	70,765	656,340	82	101,176	728,970
51	69,589	631,890	83	101,840	701,610
52	73,708	642,100	84	98,143	616,110
53	75,252	648,440	85	99,057	641,470
54	70,437	631,350	86	92,270	665,910
55	74,290	647,100	87	94,946	668,570
56	77,257	624,710	88	92,829	679,100
57	82,612	574,960	89	93,269	693,700
58	84,252	550,930			
59	86,439	582,200	1790	103,708	729,660
			91	99,383	729,800
1760	88,594	622,210	92	107,263	752,569
61	82,514	666,010	93	116,488	729,810
62	102,831	722,160	94	109,064	717,990
63	80,851	653,110			

Report on Waste Lands, 1795, pp. 202-203.

Table VIII

TALLOW CANDLES CHARGED WITH DUTY, 1711-1800
(in millions of lb.)

Year ending June 24th		Year ending June 24th		Year ending July 5th	
1711	31·4	1742	29·2	1770	43·2
12	27·2	43	30·1	71	42·7
13	31·6	44	32·1	72	42·1
14	29·2	45	32·7	73	42·2
15	30·2	46	33·5	74	43·0
16	30·6	47	34·7	75	44·1
17	30·8	48	35·0	76	46·3
18	31·7	49	34·7	77	47·3
19	33·1			78	47·9
		1750	35·0	79	46·8
1720	32·2	51	37·2		
21	33·3	52	37·6	1780	50·5
22	34·2			81	49·8
23	36·0			82	50·8
24	35·6			83	48·4
25	35·0	Year ending July 5th		84	49·6
26	35·5	1753	38·6	85	46·1
27	36·0	54	36·9	86	47·9
28	33·9	55	34·9	87	47·7
29	32·0	56	37·8	88	50·5
		57	37·1	89	51·5
1730	31·2	58	36·3		
31	32·7	59	38·2	1790	52·0
32	33·7			91	54·4
33	34·3	1760	40·3	92	54·9
34	36·0	61	41·4	93	59·1
35	37·6	62	43·2	94	59·4
36	38·1	63	42·7	95	58·1
37	36·9	64	43·3	96	56·1
38	37·2	65	42·3	97	58·7
39	37·2	66	40·2	98	61·1
		67	40·1	99	64·4
1740	34·4	68	41·8		
41	29·2	69	41·8	1800	61·7

Customs Library, *Excise Revenue Accounts, 1662-1827.*

Table IX
SOAP BROUGHT TO DUTY, 1713-1800
(in millions of lb.)

Year ending June 24th		Year ending June 24th		Year ending July 5th	
1713	24·4	1743	25·3	1771	31·1
14	24·7	44	26·5	72	30·6
15	24·6	45	26·2	73	30·6
16	23·9	46	26·5	74	31·7
17	24·9	47	27·2	75	30·8
18	25·1	48	27·0	76	31·2
19	25·7	49	27·6	77	34·8
				78	33·7
1720	26·2	1750	28·4	79	34·2
21	25·4	51	30·0		
22	26·3	52	29·8	1780	37·1
23	27·3			81	36·8
24	27·8	Year ending July 5th		82	40·9
25	26·9	1753	29·8	83	30·8
26	27·0	54	29·1	84	35·8
27	26·7	55	28·9	85	35·9
28	25·9	56	28·5	86	36·7
29	24·9	57	28·1	87	34·9
		58	27·9	88	37·6
1730	25·0	59	27·9	89	38·5
31	26·5				
32	26·6	1760	29·4	1790	40·1
33	27·7	61	29·3	91	41·9
34	26·7	62	29·4	92	42·9
35	27·5	63	29·6	93	43·0
36	27·7	64	29·9	94	47·4
37	27·7	65	29·9	95	46·8
38	27·3	66	30·0	96	46·2
39	27·5	67	29·5	97	46·2
		68	30·2	98	50·1
1740	26·2	69	30·6	99	50·6
41	24·3				
42	24·7	1770	30·8	1800	49·2

Customs Library, *Excise Revenue Accounts, 1662-1807.*

Table X

PRINTED GOODS CHARGED WITH DUTY, 1713-1800
(in million yards)

Year ending June 24th		Year ending June 24th		Year ending July 5th	
1713	2·03	1743	3·06	1771	8·74
14	2·58	44	3·04	72	9·17
15	1·84	45	2·63	73	7·53
16	2·50	46	2·73	74	8·20
17	2·65	47	3·53	75	8·16
18	2·69	48	3·22	76	8·24
19	2·84	49	4·00	77	9·02
				78	8·25
1720	1·67	1750	4·42	79	7·80
21	1·05	51	4·22		
22	1·59	52	4·21	1780	8·33
23	3·06			81	10·16
24	2·89	Year ending July 5th		82	9·61
25	2·76	1753	4·23	83	10·08
26	2·90	54	4·39	84	11·18
27	2·86	55	4·93	85	14·11
28	2·22	56	4·21	86	13·53
29	2·68	57	4·18	87	15·13
		58	5·14	88	14·60
1730	2·28	59	5·70	89	14·15
31	2·12				
32	2·43	1760	6·36	1790	16·78
33	2·92	61	6·88	91	19·65
34	2·79	62	5·62	92	21·72
35	3·00	63	5·89	93	21·14
36	2·63	64	6·63	94	20·50
37	3·06	65	6·42	95	24·05
38	3·15	66	7·16	96	30·06
39	3·22	67	7·17	97	27·20
		68	7·69	98	28·29
1740	3·12	69	9·35	99	32·18
41	3·03				
42	2·77	1770	8·72	1800	34·13

Customs Library, *Excise Revenue Accounts, 1662-1827*.

Table XI

PRODUCTION OF BROAD CLOTHS IN THE WEST RIDING OF YORKSHIRE, 1727-1800
(to the nearest thousand cloths)

Year ending March 25th		Year ending April 5th	
1727	29·0	64	54·9
28	25·2	65	54·7
29	29·6	66	72·6*
1730	31·6	67	102·4
31	33·6	68	90·0
32	35·5	69	92·5
33	34·6		
34	31·1	1770	93·1
35	31·7	71	92·8
36	38·9	72	112·4
37	42·3	73	120·2
38	42·4	74	87·2
39	43·1	75	95·9
		76	99·7
1740	41·4	77	107·8
41	46·4	78	132·5
42	45·0	79	110·9
43	45·2		
44	54·6	1780	94·6
45	50·5	81	102·0
46	56·6	82	112·5
47	62·5	83	131·1
48	60·8	84	138·0
49	60·7	85	157·3
		86	158·8
1750	60·4	87	155·7
51	61·0	88	139·4
52	60·7	89	154·1
Year ending April 5th			
1753	55·4	1790	172·6
54	56·1	91	187·6
55	57·1	92	214·9
56	33·6	93	190·3
57	55·8	94	191·0
58	60·4	95	251·0
59	51·9	96	246·8
		97	229·3
1760	49·4	98	224·2
61	48·9	99	272·8
62	48·6		
1763	48·0	1800	285·9

* The figures for the years before 1768 are unreliable. The increase after 1765 is partly to be explained by improved inspections.

David Macpherson, *Annals of Commerce* (1800), vol. iv.

Table XII

PRODUCTION OF NARROW CLOTHS IN THE WEST RIDING OF YORKSHIRE, 1738-1800
(to the nearest thousand cloths)

Year ending Jan. 20th		Year ending Jan. 31st	
1738	14·5	1770	85·4
39	58·8	71	90·0
		72	95·6
1740	58·6	73	89·8
41	61·1	74	88·4
42	62·8	75	96·8
43	63·5	76	95·6
44	63·1	77	95·8
45	63·4	78	101·6
46	68·8	79	93·2
47	68·4		
48	68·1		
49	68·9	1780	87·4
		81	98·8
1750	78·1	82	96·8
51	74·0	83	108·6
52	72·4	84	115·6
Year ending Jan. 31st		85	116·0
53	71·6	86	123·0
54	72·4	87	128·8
55	76·3	88	132·2
56	79·3	89	145·4
57	77·1		
58	66·4		
59	65·5	1790	140·4
		91	155·4
1760	69·6	92	190·4
61	75·5	93	150·6
62	73·0	94	130·4
63	72·1	95	155·0
64	79·5	96	151·6
65	77·4	97	156·8
66	78·9	98	148·6
67	78·8	99	180·2
68	74·5		
69	87·8	1800	169·2

David Macpherson, *Annals of Commerce* (1800), vol. iv.
See note to Table XI.

Table XIII
THE AVERAGE PRICE OF BANK STOCK 1700-1800, OF 3% ANNUITIES 1726-57 AND OF 3% CONSOLIDATED STOCK 1758-1800.

September	Bank Stock £	3% £	September	Bank Stock £	3% £
1700	136	—	1750	135	100
01	116	—	51	142	99
02	125	—	52	144	105
03	135	—	53	137	104
04	125	—	54	132	104
05	92	—	55	123	90
06	81	—	56	117	88
07	111	—	57	120	91
08	125	—	58	118	89
09	131	—	59	112	81
1710	$113\frac{1}{2}$	—	1760	111	82
11	$104\frac{1}{2}$	—	61	111	74
12	114	—	62	109	80
13	127	—	63	116	84
14	132	—	64	122	83
15	128	—	65	136	89
16	138	—	66	139	87
17	150	—	67	152	88
18	146	—	68	167	89
19	144	—	69	$168\frac{2}{3}$	88
1720	216	—	1770	137	78
21	130	—	71	114	87
22	$115\frac{1}{2}$	—	72	148	89
23	122	—	73	143	87
24	130	—	74	142	88
25	135	—	75	141	89
26	$127\frac{1}{4}$	—	76	137	83
27	133	84	77	130	78
28	137	—	78	114	64
29	$139\frac{1}{2}$	93	79	111	61
1730	$144\frac{1}{2}$	91	1780	114	63
31	148	96	81	110	56
32	152	99	82	114	57
33	143	97	83	127	66
34	140	94	84	111	54
35	140	94	85	122	65
36	151	105	86	151	74
37	145	106	87	148	69
38	145	105	88	172	74
39	134	98	89	189	80
1740	144	100	1790	181	77
41	141	99	91	200	89
42	143	100	92	200	90
43	148	101	93	172	74
44	147	93	94	164	66
45	141	85	95	169	69
46	135	88	96	139	56
47	126	82	97	130	50
48	128	88	98	131	50
49	139	101	99	171	64
			1800	171	65

From 1700 to 1720 the figures have been obtained from prices given in J. Castaing, *Course of the Exchange*; 1721 to 1730 from those given in J. E. T. Rogers, *A History of Agriculture and Prices in England*, Vol. VII, Pt. II. The figures for the remaining years are those given in Sir John Sinclair, *The History of the Public Revenue of the British Empire* (1803), Vol. II, Appendix II.

Table XIV
TOTAL ENGLISH EXPORTS (EXCLUSIVE OF SPECIE) AND TOTAL IMPORTS, 1700-1800

Year ending December 25th	Exports m£	Imports m£	Year ending December 25th	Exports m£	Imports m£
1700	6·5	6·0	1751	12·4	7·9
01	6·9	5·9	52	11·7	7·9
02	4·8	4·2			
03	6·2	4·5	*Year ending January 5th*		
04	6·2	5·4	1753	12·2	8·6
05	5·3	4·0	54	11·8	8·1
06	6·2	4·1	55	11·1	8·8
07	6·4	4·3	56	11·7	8·0
08	6·6	4·7	57	12·3	9·3
09	5·9	4·5	58	12·6	8·4
			59	13·9	8·9
1710	6·3	4·0			
11	6·0	4·7	1760	14·7	9·8
12	6·9	4·5	61	15·0	9·5
13	6·9	5·8	62	13·75	8·9
14	8·0	5·9	63	14·7	11·2
15	6·9	5·6	64	16·3	10·4
16	7·0	5·8	65	14·6	11·0
17	8·0	6·3	66	14·1	11·5
18	5·4	6·7	67	13·9	12·1
19	6·8	5·4	68	15·1	11·9
			69	13·4	11·9
1720	6·9	6·1			
21	7·2	5·9	1770	14·3	12·2
22	8·3	6·4	71	17·1	12·8
23	7·4	6·5	72	16·2	13·3
24	7·6	7·4	73	14·8	11·4
25	8·5	7·1	74	15·9	13·3
26	7·7	6·7	75	15·2	13·6
27	7·3	6·8	76	13·7	11·7
28	8·7	7·6	77	12·7	11·8
29	8·2	7·5	78	11·6	10·3
			79	12·7	10·7
1730	8·5	7·8	1780	12·6	10·8
31	7·9	7·0	81	10·6	11·9
32	8·9	7·1	82	13·4	9·5
33	8·8	8·0	83	13·9	12·1
34	8·3	7·1	84	14·2	14·1
35	9·3	8·2	85	15·1	14·9
36	9·7	7·3	86	15·4	14·6
37	10·1	7·1	87	15·8	16·3
38	10·2	7·4	88	16·3	16·6
39	8·8	7·8	89	18·2	16·4
1740	8·2	6·7	1790	18·9	17·4
41	9·6	7·9	91	21·4	17·7
42	9·6	6·9	92	23·7	17·9
43	11·3	7·8	93	19·4	17·8
44	9·2	6·4	94	25·7	20·8
45	9·1	7·8	95	26·3	21·5
46	10·8	6·2	96	29·2	21·5
47	9·8	7·1	97	27·7	19·5
48	11·1	8·1	98	31·9	26·0
49	12·7	7·9	99	34·1	24·5
1750	12·7	7·8	1800	40·8	28·4

The figures of Exports include Re-exports.

Table XV

RATE OF STERLING EXCHANGE ON HAMBURG, 1700-1800
(in Schillings and Grotes)

(The first January quotation is given for each year)

Year	Rate	Year	Rate	Year	Rate
1700	34·2	1734	35·5	1768	34·11
01	34·6	35	35·3	69	33·2
02	35·4	36	35·4		
03	33·3	37	35·2	1770	33·2
04	32·5	38	35	71	33·8
05	32·10	39	34·6	72	32·7
06	33			73	34
07	34	1740	34·6	74	34·9
08	33·4	41	34	75	34·3
09	33·4	42	33·11	76	34·1
		43	33·10	77	33·2
1710	32·7	44	33·9	78	32·4
11	33·11	45	36·3	79	35·6
12	33·7	46	36·7		
13	34·3	47	35·7	1780	34·6
14	34·7	48	35·1	81	34·1
15	35·1	49	34·6	82	31·9
16	34·7			83	32·7
17	34	1750	33·10	84	33·6
18	33·5	51	33·5	85	34·10
19	34·1	52	33·7	86	34·10
		53	33·2	87	34·5
1720	34·3	54	33·3	88	35·1
21	32·10	55	33·8	89	34·10
22	34·9	56	34·9		
23	34·9	57	36·7	1790	35
24	33·10	58	35·8	91	35·6
25	33·9	59	35·8	92	34·6
26	33·10			93	35·4
27	35·2	1760	36·4	94	35·9
28	34·4	61	32	95	34·6
29	33·5	62	32·11	96	32·7
		63	34·3	97	35·6
1730	32·11	64	34·5	98	38·2
31	33·5	65	35·1	99	37·7
32	34·1	66	34·6		
33	34·2	67	35·6	1800	32

The rates from 1700 to 1720 and from 1747 to 1800 (12 Grotes to a schilling) are taken from *The Course of the Exchange* printed by J. Castaing and kept in the Library of the Stock Exchange London: 1721 to 1746 from *British Parliamentary Papers*, 1810-11, vol. X.

Table XVI
BANKRUPTCIES, 1732-1800

Year ending September 30th		Year ending September 30th	
1732	164	1767	334
33	170	68	303
34	207	69	309
35	218		
36	212	1770	393
37	222	71	345
38	210	72	484
39	265	73	556
		74	338
1740	291	75	332
41	255	76	388
42	—	77	489
43	185	78	662
44	162	79	575
45	175		
46	167	1780	454
47	179	81	381
48	201	82	411
49	207	83	540
		84	534
1750	211	85	383
51	172	86	509
52	170	87	487
53	197	88	697
54	243	89	560
55	231		
56	236	1790	574
57	250	91	603
58	278	92	609
59	280	93	1256
		94	857
1760	216	95	731
61	178	96	720
62	188	97	905
63	214	98	767
64	284	99	512
65	223		
1766	288	1880	727

London Gazette, 1738-41 and 1786-1800.
The Gentleman's Magazine, 1732-37 and 1743-85.

Index

ADAM BROTHERS, 25
Agriculture, Ch. II *passim*.
Anglesey, copper-mining, 92
Animal husbandry, 51, 52
Apprenticeship, 5, 9, 15, 16, 137, 139, 148, 223-4, 229
Arboriculture, 43
Arbuthnot, John, 43, 87
Arkwright, Richard, 107, 117, 214, 228
Arsenal, Royal, 113
Assignats, 187
Assize of Bread, 56
Attorneys, 182

BACON, ANTHONY, 136, 138
Balance, of trade, 192; of payments, 61, 151
Bank Notes, 178-9
Bank of England, 178-9
Banks, private, 178-85; London, 179-80; country, 180-5; rural, 184-5
Barings, the, 137-8
Barnard, Sir John, 111, 133, 134, 137, 172, 196
Beer, 59
Berkeley, Bishop, 213
Bills, inland, 185-7; of exchange, 188-9
Bills of Mortality, London, 4n.
Birmingham, 13, 97, 104, 232
Bonds of service, 211
Boulton, Matthew, 173, 182
Boulton and Watt, 211, 232
Braund, Samuel, 131; William, 138
Bridgewater, Duke of, 74, 77, 82, 206
Brindley, James, 74, 76
Bristol, 12, 26, 89, 92, 142, 145
Bristow, John, 136, 196
Brokers, bill, 135-6; insurance, 133; ships', 132
Building industry, 97
Bullion dealers, 190
Burrell, Peter, 136, 137, 196
Business, size of, 98-100; growth of, 113-16

CANALS, 74-7; Bridgewater, 74, 77, 82, 88; Grand Trunk, 75; Grand Junction, 75; Sankey, 74, 81, 83; Weaver, 83; benefits of, 84-7; chronology, 84; rates and tolls, 88
Candles, 59-60
Capital, fixed and circulating, 100; import of, 127
Carron Iron Co., 120

Cattle, sales at Smithfield, 53; weight of, 51; disease, 53
Charter-masters, 121-2
Cheesemongers, 66
Class structure, 20
Clay, 92
Cloth Halls, 63-4, 97
Clothiers, 98-102
Coal, Exchange, 64, 66; trade, 70, 72, 92, 93
Coalbrookdale, 67, 178, 209, 213, 220
Coastal trade, 70-2
Coin, 167-77; shortage of, 207
Colquhoun, Peter, 72
Combinations of traders, 66; of manufacturers, 122-4; of workers, 226-31
Companies, Chartered, 130
Convoy Duty, 152
Corn Laws, 48-50
Cornwall, mining, 92-3, 118
Cort, Henry, 117
Cost of living, 94, 233-4
Cottagers and squatters, 47
Counterfeiters, 174-6
Credit, 206, 209-10
Cromford, 117, 214
Crowley, Ambrose, 93, 94, 114, 120, 121, 212

DARBY, ABRAHAM, 13, 67, 92, 213
Dearths. *See* Harvests.
Debasement, of coinage, 175
Debtors and creditors, 206
Debts, small, 207, 209-11
Deductions from wages, 218
Defoe, Daniel, 5, 33, 38, 99, 113, 122
Diet, 8
Dissenters, 3
Distilling, 57
Division of labour, 103-4
Docks and harbours, 142
Drinking habits, 217
Dutch investment, 193

EAST INDIA COMPANY, 160-1, 170
Eden Treaty, 166
Egertons, the, 139
Embezzlement, of materials, 208-11
Emigration, 11
Enclosure, 38-48; number of Acts of, 40
Entail, 24
Estates, landed, 34-5
Exchanges, foreign, 188-96; brokers, 186
Excise duties, 115

255

FACTORS, 66-7, 135
Fairs and markets, 63-4
Family business, 118
Finney, Samuel, 214-15; Peter Davenport, 69, 181
Freehold, 36
Fulling mills, 95
Furnese, Sir Henry, 136, 195

GAMBLING, 24-6
Gideon, Simon, 137
Gin, 6-7, 57-8
Gold, overvaluation of, 170
Gold Standard, 177
Goldney, Thomas, 67, 178
Goldsmid, Aaron, 136
Goldsmiths, 179
Gore, John, 136, 140, 196
Gott, Benjamin, 139
Government establishments, 113-14
Greg, Samuel, 216
Guineas, value of, 168, 172
Gurneys, the, 184

HARRIS, JOSEPH, 169, 176
Harvests, effects of yields of, 4, 61-2, 192, 199; and riots, 227
Highways, 78-81
Hindley, Henry, 133, 135, 146, 155
Holdings, agricultural, 44, 46
Holidays, 204
Hops, 58
Horsehay Company, 71
Horse-power, 110, 116
Horses, 55, 86, 87, 116
Hours of work, 212, 230
Howlett, John, 15
Hunter, John and William, 8
Huntsman, John, 139

IMMIGRATION, 10, 127-8
Import duties, 125-6
Income, distribution of, 22, 199
Index numbers, 197
Industrial Revolution, 125
Infant mortality, 9
Innovations, effects, 108-13
Insurance, 25, 132-4; brokers, 133
Interest, rate of, 26-28; and enclosures, 41; land values, 45; raising of cattle, 52; storage of grain, 67; turnpikes and canals, 84, 89; invention, 105; patents, 108; trade, 137
Invention, 104-8; effect on labour, 108-9; on capital, 110-11
Irish, 10, 32n., 175, 222, 227-8, 235
Iron industry, location, 92, 95

JANSSEN, SIR THEODORE, 191
Jenny, the, 116

Jews, the, 172, 190, 222
Joint-stock companies, 18, 75, 119, 130

KING, GREGORY, 2, 11, 21, 22, 23
Kinship, 18

LABOUR, Ch. VII *passim*; on canals 77; statute, 78
Lancashire, 13, 16, 185, 232
Land Tax, 36, 37
Leases, 37
Leather, 60
Liverpool, 17, 74, 81, 89, 92, 138, 140, 142, 145, 185
Lloyds, 133
Localisation of industry, 91-5
Lombe, Sir Thomas, 96, 116
'Long pay', 207
Lotteries, 24-5
London, 10, 31, 64-5, 66, 69, 81, 86, 89-90, 136-7, 140, 145, 179

MACHINERY, ATTACKS ON, 228
Macpherson, David, 63, 148, 199, 210
Malt, 58
Manchester, 96, 97, 106, 116, 185, 232
Manufacturers, 21, Ch. IV *passim*.
Marsden, Thomas, 180-1
Marshall, William, 44-5
Medicine, improvements in, 8
Merchants, 21, 130-40; alien, 140
Metcalf, John, 81
Migration, internal, 13-17, 94
Milling and baking, 56-7
Mint, Royal, 167-8, 171; ratio, 168-9
Money, Ch. VI *passim*
Money scriveners, 182
Mortgages, 27-8
Mule, the, 116
Murdoch, William, 21, 107, 150

NATIONAL DEBT, 26
Navigation Acts, 145, 147, 150
Newcastle, 17. 26, 92, 227
Newcomen, Thomas, 96
Newton, Sir Isaac, 170, 171, 176, 193n.
'Nickoll, Sir John', 104n., 217n.

OLDKNOW, SAMUEL, 183, 187, 208
Open-field agriculture, 33-4
Output, industrial, 124

PAR OF EXCHANGE, 189
Partnerships, 118, 131
Patents, 106-8
Paul, Lewis, 110, 116, 211
Pawnbrokers, 210
Perquisites, 218
Philips, J. and N., 180
Pioneer economy, 232

Pinney, John, 24, 131, 139
Poor Law, 5, 14, 15, 16, 235
Population, Ch. I *passim*; geographical distribution, 12-17
Postlethwayt, Malachy, 130-1, 137, 139, 205
Press-gangs, 149
Prices of cereals, 39-40; animal products, 39, 54, 57; local variations, 86; general movements, 197-200
Priestley, Joseph, 187
Product-sharing, 118, 208
'Progressive state', 231
Public mills, 97
Public works, 81-2
Putters-out, 101-2

QUAKERS, 57

RECEIVERS OF TAXES, 181-2
Recoinage (1696-8), 170; (1733), 175; (1773-4), 167, 183
Religious persuasions, 19-20
Rent, agricultural, 37, 45-6
Restrictive practices of manufacturers, 123; of traders, 66; of workers, 223-4, 229-30
Retailers, 68-70, 180, 209-10, 216
Reynolds, William, 75, 120, 212
Rickman, John, 2
Riots, 86, 227-9
Rivers, 72-3

SALT, 92
Sankey Navigation, 74, 81
Savile, Sir George, 186
Saving, 23
Seamen, 144, 148, 149, 206, 226
Seasonal fluctuations of employment, 202-3
Servants, 5, 128, 203
Settlement, Act of, 14
Sheep, sales at Smithfield, 53; disease, 53-4
Sheffield, 13, 15, 16, 81, 97, 104, 232
Shipbuilding, 97-8, 140-1
Shipping, partnerships in, 118; statistics of, 141; distribution by ports, 144-5; losses in war, 146; foreign, 147
Ship's brokers, 132; husbands, 131
Silver coin, shortage of, 170-2; replacement by gold, 177
Sinclair, Sir John, 51
Smallpox, 8
Smith, Adam, 14, 42, 48, 52, 54, 61, 71, 110, 111, 119, 146, 148, 165, 172, 176, 192, 205, 213, 219, 220, 221, 224, 230, 231, 232
Smithfield Market, 51
Smuggling, 162-5

Soap, 60
Society of Arts, 106
Specie points, 190
Spitalfields Market, 219
Starch, 58
Stout, William, 60-1, 68, 69-70, 71, 145, 181
Strutt, Jedediah, 28-9, 139
Stubs, Peter, 21, 68, 69, 101, 102, 139; Sarah, 13
Supercargoes, 135
Suspension of cash payments, 199

TANNING, 60
Tea, imports of, 165
Telford, Thomas, 113
Temple, William, 202, 235
Terms of trade, 234
Textile trades, localisation, 94, 95
Thornton, Henry, 137
Tokens, traders', 174
Touchet, Samuel, 137, 138
Towns, growth of, 8, 15, 95-6
Townshend, Charles, 65, 201
Trade, internal, 63-70; overseas, Ch. V *passim*; with Europe, 155; Ireland, 155-6; Atlantic, 156-9; Far East, 159-60; entrepôt, 161; slave, 166
Trade unions, 226-31
Truck, 174, 202, 208
Tucker, Dean 81, 217
Turnpikes, 79, 80, 86, 87

UNDERWRITING, 132-3
Unemployment, 203
Usury laws, 27-9

VANNECK, SIR JOSHUA, 136

WAGES, 218-26, 232-4
Walker, Samuel, 75, 80, 120, 183
War and agriculture, 46; industry, 126-7; wages, 225-6
Watch-making. 103
Water and water power, 93-4, 109
Water-frame, 116-17
Watt, James, 107, 149
Wedgwood, Josiah, 74, 139, 212
Wheels, broad and narrow, 79
White, Gilbert, 8
Whitworth, Richard, 85
Wilkinson, John, 174, 175, 183, 187
Williams, Thomas, 174
Wool, 59
Wool-combers, 230
Workhouses, 114

YOUNG, ARTHUR, 16, 44, 51, 86, 202